Bird-Watching in London

BIRD-WATCHING IN LONDON

A HISTORICAL PERSPECTIVE

by

E. M. Nicholson

LONDON NATURAL HISTORY SOCIETY

1995

Text copyright © London Natural History Society 1995

First published 1995

LNHS

Published by the
London Natural History Society
Registered Charity No. 206228
c/o The Natural History Museum
Cromwell Road, London SW7 5BD

Cover printing generously donated by
John Parfitt, fine art printing consultant

Cover design and map by Ken Osborne

Front cover illustration of gulls over
the Round Pond in Kensington Gardens
by Charles Tunnicliffe, reproduced by permission
from *Bird Life in the Royal Parks 1961–62* (HMSO)

The photograph on the back cover by W A Smith
shows the Round Pond in 1928

Extracts may be reproduced for educational,
scientific or non-profit-making purposes without permission
provided full acknowledgement of the source is given

Printed and bound in Great Britain
by Antony Rowe Ltd, Chippenham, Wiltshire,
from camera-ready copy prepared
and edited by Mike Earp

British Library Cataloguing-in-Publication Data

A catalogue record for this book is available from
the British Library

ISBN 0 901009 05 9

CONTENTS

	Editor's Preface and Acknowledgements	vii
	Preface	xi
Chapter 1	The London Ducks	1
Chapter 2	The London Gulls	20
Chapter 3	The Flylines over London	46
Chapter 4	Kensington Gardens and Hyde Park	73
Chapter 5	A Bird Census of Kensington Gardens	116
Chapter 6	Bird Protection in London	136
Appendix	The Birds of Hyde Park and Kensington Gardens, by R. F. Sanderson	147
	Index	199

Editor's Preface and Acknowledgements

The bulk of this work is based on material Max Nicholson prepared in the mid 1920s for a popular book to be entitled *Bird-Watching in London*. That material, which deals largely with his own detailed observations on the birds of Kensington Gardens and Hyde Park, was never published, though some of it has appeared in shortened form over the years in various of his publications.[1] Mr Nicholson, an Honorary Vice-President of the London Natural History Society since 1964, recently offered the work to the Society to publish if it wished (as described in the following letters). I have had the great pleasure of preparing this interesting and historically important material for publication.

Max has strong London connections, having lived in the capital most of his life. He was a member of the Committee on Bird Sanctuaries in the Royal Parks from its re-creation in 1947 until its demise in 1979. Most of the present book was completed by 1927, when Max was still only 22 years of age, though by then his first path-breaking book, *Birds in England*, had already been published; Chapter 6, on Bird Protection in London, was written a couple of years later. *Bird-Watching in London* is in the genre of W. H. Hudson's classic *Birds in London* (1898) and should be viewed in that light; it is not a scientific treatise. Hudson was a contemporary of Max's, whose concern about the establishment of sanctuaries for wildlife in the parks he shared and whose works he read when young to help equip himself as a writer and journalist.

Some new material has been added, as described below, but Max's original title *Bird-Watching in London* has been retained, though with the addition of the subtitle *a historical perspective*, and no attempt has been made to update or change the character of the original text. Instead, Roy Sanderson—one of the last Official Observers of Hyde Park and Kensington Gardens, and noted for his long-term census work in the parks—has kindly provided a full account of the bird-life of the two parks, based on his own extensive observations plus the published records of others, which is included as an appendix. This brings the picture up to date for the sites which form the focus for Max Nicholson's account and sets his observations in context. Publication in this form has permitted a lengthier treatment than would have been possible if the account had been included in the *London Bird Report*.

Max and Roy have also consented to the inclusion, at the end of Chapter 5, of their previously unpublished account of the 1975 census of

---

1. The MS of chapters on The London Sparrows, Inner London and Outer London was stolen in a suitcase from EMN's car in Liverpool around 1927.

Editor's Preface and Acknowledgements

Kensington Gardens. This book is published on the eve of the twentieth anniversary of that census and the seventieth of Max's first census in Kensington Gardens. Except in that account and the appendix, which use current names and taxonomy, birds' names are as given in Max's original manuscript, with lowercase initials. They are indexed under modern equivalents with cross-references where necessary. Measurements in the main text have, similarly, not been converted to their metric equivalents.

I have added a number of editorial footnotes. The intention of these is to provide additional references, and comments made with the advantage of hindsight, for the benefit of today's reader. These include many helpful comments made by Bunny Teagle, who kindly read through a complete draft. Evelyn Brown, Dr Colin Bowlt (LNHS President), Dr Paul Cornelius (LNHS Archivist) and Keith Hyatt (editor of *The London Naturalist*) also provided helpful comments on various drafts.

Gina Douglas—a former colleague of Max's on the International Biological Programme and now Librarian at the Linnean Society of London—very kindly transcribed and typed up a chapter and a half of MS material. I am extremely grateful to her for agreeing to do this and for doing it so well and so quickly, also for allowing me to refer to a number of old journals in the Linnean Society Library.

Colin Bowlt, Doug Boyd, Paul Cornelius, Ann Gooderham, Mark Hardwick, Ken Osborne and Bunny Teagle are to be thanked for providing valuable comments on the appendix by Roy Sanderson. The Park Manager, Jennifer Adams, kindly provided Roy with data on tree losses in Hyde Park and Kensington Gardens. I am extremely grateful to Colin J. Raymond for typing up the original draft of Roy's contribution and supplying it on disk and to Ken Osborne for producing the map of the two parks which appears on pages 2 and 3 and for designing the cover and arranging for its sponsorship. The front cover illustration by Charles Tunnicliffe is included courtesy of Ken; Paul Cornelius found the photo of the Round Pond, which appears on the back cover, in the Society's archive.

The publication of this work in its present form has been overseen and supported by the LNHS's Publications Working Group under its Chairman, Tony Barrett.

Mike Earp

Editor's Preface and Acknowledgements

BIRD-WATCHING IN LONDON

Extract from Note for the LNHS

22nd February 1994

By Max Nicholson

During 1924, 1925 and 1926 I spent much time conducting intensive observations on the birds of inner London and writing them up systematically. Some of my notes were used for articles published in *Discovery*, *The Field*, *The Nineteenth Century and After* and other journals and newspapers. Although these were well received, when I used them to prepare a book enquiries at all the main booksellers such as Hatchards, Sotherans, Harrods, and Truslove and Hanson elicited a unanimous view that there was no market for such a work, and it has therefore reposed in an unopened packet for more than sixty years. Meanwhile interest in London birds has grown and these quite vivid facts about how it was around seventy years ago have acquired some historic value.

In view of the above, and of the appreciation I feel for the London Natural History Society, I am minded to present the typescript with the copyright of this work to the London Natural History Society unconditionally, either to be held for reference in the archives or if so desired to be suitably published for the benefit of the Society, in a form approved by me as of acceptable standard, any proceeds accruing to the LNHS.

Meanwhile I am passing the typescripts of the five extant chapters to Ruth Day (my letter to her of 22nd February 1993[2] refers).

Max Nicholson

---

2.   ED: See the extract reproduced overleaf.

Editor's Preface and Acknowledgements

Extract from letter of 22nd February 1993 to Ruth Day, then LNHS President

... soon after my first book *Birds in England*[3] was published in 1926, I virtually completed a book of my very extensive observations in London during 1925–27 entitled *Bird-Watching in London*. Unfortunately, when the prospective publishers canvassed the London bookshops, they almost unanimously responded that there would be practically no buyers for such an esoteric subject and the MS went into a bottom drawer from which I have only just unearthed it. It records many facts about the different species in different parts of London some 66–68 years back. If the LNHS would like to take it over I would be happy to donate it and to sign papers transferring the copyright and all other rights in it, so that at some time it could if desired be published in whole or in part by or for the Society. If you will in due course let me have a separate answer on this, it is ready to be handed over, still in good shape, although I wonder how much further exposure to the elements its flimsy typing paper will withstand!

---

3. ED: The following passage from *Birds in England* (pages 115–116) also relates to London:

    '... The vast expansion of modern London has driven out all the original inhabitants from the land over which it is strewn. For some time after the growth became abnormally rapid all central London was left temporarily birdless—the old fauna was driven out and nothing except sparrows took its place. But, arid as it seemed, London did offer a living to other creatures besides pigeons and sparrows. The last generation has seen the acquisition—which is still in progress—of a peculiar metropolitan fauna, a strange menagerie of very diverse forms with only the tameness of their manner and the prevailing sootiness of their dress in common. The black rat has revived his fortunes, long eclipsed, by acquiring, among other tricks, the knack of travelling by the telegraphs, and with the alien grey squirrel has made himself a characteristically metropolitan beast. Of birds, the starling, carrion crow and woodpigeon in the parks, the moorhen, mallard and dabchick on the ponds, and a few species of small birds have succeeded in making themselves widespread residents. The winter migrants are numerous—starlings, black-headed gulls and tufted duck are the most conspicuous. Not only sanctuaries and goodwill on our part, but realisations of new possibilities on the side of the birds, will ensure that the next town-living generation will be better provided than its predecessors with town-living birds. Some provision for smoke abatement, which can hardly be deferred much longer, will improve the plumage of the present wild creatures in sooty districts and make these tenable for others.'

Preface

We are apt to look upon London as a town inhabited chiefly by sparrows, distinguished from the sparrows of the rest of the world only by a more dingy plumage and by their fabulous abundance. Certainly there is not a great variety of metropolitan species, and there are vast barren tracts of slate and brick more utterly devoid of wild life than any other part of the land surface of the three kingdoms—as dead and naked as if they had been sterilised in a laboratory, but not nearly as clean. For all that, London is ornithologically more important, and certainly more interesting to the ordinary birdlover, than the great majority of perfectly rural neighbourhoods. Although there are fewer species, and generally fewer birds, they live in closer contact with us and constantly give opportunities for intimate observation which come to a country observer only once in a blue moon. For it is only in this least natural place in the whole kingdom that man occupies at this day his natural place in the scheme of things. Elsewhere he and his civilisation are entrenched against nature: he is feared and avoided—on the marshes duck rise in flocks as he shows himself within half a mile of them and the beasts become so timid and nocturnal that there is no country in the world where one sees so few wild mammals alive. But here birds look upon him as a harmless or beneficent creature: even to the wild pochards and black-headed gulls which come down in winter from the north he is a being no more to be avoided than a swan or a sheep. If he brings them offerings of food they take it boldly and quite trustfully: otherwise they do not much concern themselves about his presence. Dogs demand far more of their attention: if it is only a man they go on with whatever they are doing. Here in fact he receives freely in return for his good behaviour the privilege which ornithologists have otherwise to seek partially and unsatisfactorily by concealing themselves in hides, or by lying very still and hoping not to be noticed—the power of watching birds openly at close range without making them uneasy. That is the great thing: a breath of suspicion or fear in the performers is fatal to the successful observation of all the most secret and significant ceremonies in which birds take part. In addition to these, there are many peculiar habits observed only in the metropolis, or not commonly observed away from it. The lesser variety of birds, their greater dependency on man, and the intense rivalry between the competing species produces a state of things not only far more exaggerated than the normal invisible struggle for existence but much more easily understood. The parks moreover—at any rate the Royal Parks of mid London—are so perfectly isolated from the rest of the green world that they form a scientifically exact barometer for recording seasonal movements. When birds are met with which are not normally resident it is a clear proof of at any rate local migration. But London is not all parks. The dreary wastes of poverty-stricken dwellings and the hardly less barren deserts of respectability are frankly not worth much notice. Sometimes on migration a water-rail or some other unexpected visitor may be found with a broken wing, dropped like a portent out of the skies, but otherwise such birds as are able to put up with these conditions at all are common birds overwhelmed and made dingy and banal by their dingy and banal surroundings. Yet besides the parks and the dwellings there is still something else. There is central London, with the hordes of starlings (which impress the visitor to the British Museum on a winter afternoon more deeply than anything he will find inside it) and the pigeons, and the Embankment off which the large maritime gulls, following the highway of the Thames, halt and remain in security. What more there is depends entirely on what area is accepted as being London. Even within the county boundary there are Hampstead Heath and Ken Wood and Parliament Hill, a good deal of Wimbledon Common and part of Richmond Park, green open suburbs where the cuckoo still makes his voice heard, and other places notable for their birds. But the arbitrary county boundary is no true limit: in this book what counts as London is taken to be the territory which London has for one purpose or another annexed or altered in character. The large reservoirs about Staines, for example, are clearly London, for they owe their existence simply to the immensity of a metropolitan thirst. So are Kew Gardens London, and Hampton Court and all the suburbs; but the fields about Hanwell are not London, nor the country between Edgware and Elstree, for, though they are nearer to Charing Cross than many which have been included, London has not yet laid hands on them so as to change at all appreciably the character of their wild life. The principle may be criticised as artificial: so are boundaries. In a book which set out to give a history of London birds some other plan would certainly be desirable. But here there is no such ambition: I have not attempted

Preface

to give an account of anything outside the scope of my own very limited observations, made, some systematically, some at haphazard, during 1924, 1925 and 1926. I know too little of the birds of London, and my London experience is much too brief, to entitle me to undertake anything further, especially since there are at least three[4] living observers, whose names are well known, possessing infinitely greater qualifications and minds stored with a generation's experience of the metropolitan bird-life. It would all the same be wrong to neglect the opportunity of calling attention to the fact that London, which possesses so many birdlovers and such a notable avifauna, has less serious bird literature, and far worse bird literature, than almost any important area in the kingdom. There is no county history, in print or out of print, to compare even with such a work as Borrer's *Birds of Sussex*[5]—the Victoria County History of London seems to be the only one of its series where no account of the birds is given—and among other books W. H. Hudson's *Birds in London*, by no means one of his best, is apparently the only work still current[6] which has any merit at all. This, surely, is a very scandalous neglect; there is a conspicuous need for a worthy history of the Birds of London.[7] Even to imagine such a book is to realise the shortcomings of this one, in which nothing more is aimed at than to give some sense of the richness of the metropolitan bird-life, and of the peculiar pleasure that comes from bird-watching where birds are almost the only stray particles of nature to be found.

---

4. EMN: One whom I had in mind, Mr Harold Russell, has died since this was written. He was an excellent observer, keen and well-seasoned: there was no one who knew the birds of Hyde Park and Kensington Gardens so intimately, or who did so much towards their welfare, for it was at his initiative that the London Bird Sanctuaries were set up.
5. ED: W. Borrer 1891, *The Birds of Sussex*.
6. ED: It was published in 1898.
7. ED: A need first addressed by W. E. Glegg's *A History of the Birds of Middlesex* (1935). Earlier books on London's birds included J. E. Harting's *The Birds of Middlesex* (1886), H. K. Swann's *The Birds of London* (1893), T. Digby Pigott's *London Birds and Other Sketches* (1884, revised 1892 & 1902) and Charles Dixon's *The Bird-Life of London* (1909).

Preface

In description I have used as far as possible the exact words in which things seen were noted at the time. Many of the incidents have already been described in articles, and I have to thank the editors of *The Field*, the *Evening News*, the *Daily News* and *English Life* for permission to incorporate them in their proper places. A summary of most of the observations included in Chapter 4 appeared in *The Nineteenth Century and After* for December 1925[8] as *The Birds of Kensington Gardens*, and a partial summary of *The Flylines over London* in the *New Statesman* in March 1926, while much of the material in Chapter 5 appeared in the August 1926 issue of *Discovery*.[9] To the editors of these I am also indebted.

But this book owes most of all to the help of my brother, Basil D. Nicholson,[10] freely given in the field and afterwards. Particularly the difficult work involved in the bird census of Kensington Gardens, and in attempting to map flylines over London, would have been beyond my power without him, and more lately his sounder judgement has purged the manuscript of some of its most conspicuous blemishes.

E.M.N.

HASLEMERE
New Year's Day 1927

---

8. ED: vol 98(586): 922–932.

9. ED: vol 7(80): 281–285.

10. ED: Lived 1907–1953.

Chapter I

The London Ducks

In appearance the London **mallard** is identical with mallards of the rest of Britain, or even of Asia and North America, but he is nevertheless a different bird. They are shy and rise at long gunshot; he is as trustful of man as a sparrow. They feed chiefly by night and spend the daytime in repose—up the Thames Valley there are watercress beds where red lights have to be kept burning all through the hours of darkness to frighten away the night-flying wild ducks, whose descent would play havoc with the floating crops; in London, where he depends largely on the food we offer him, the mallard appears as much a bird of the day as the wintering gulls, which retire from the metropolis every sunset and return at dawn. The contrast here is actually not so great as it seems, for even in central London he often dozes through the day and wakens into activity at night. He is distinguished also by a series of peculiar habits which have been adopted one by one to cope with the peculiar circumstances of a metropolitan existence, and for the field-naturalist these cockneyisms have a greater fascination than the most striking variations in plumage. They reveal the bird's mind and personality, a much more interesting thing than its carcass.

The mallard's year begins early. Our summer is his dead season. From about the end of June more or less until September he lurks in dense cover, as dingy and dishevelled as a pullet in a backyard, and as little capable of flight. But by September, or at latest October, he again assumes his metallic splendour, receives new wings to last him till the succeeding midsummer, develops that expressive masculine curl of the tail-feathers which is the emblem of the mature drake and, having thus recovered his self-respect after the chastening episode of the moult, begins once more to pay attention to the other sex. Anyone who often passes the waters of the Royal Parks in late autumn or winter must have noticed something of the ceremony of courtship, even if he failed to understand its meaning, for the stiff and jerky gallantry of the excited birds seems as comic to our eyes as it must seem romantic and deadly serious to theirs. It is at that early season that the balance of power between the sexes is most evenly poised. The

Kensington Gardens

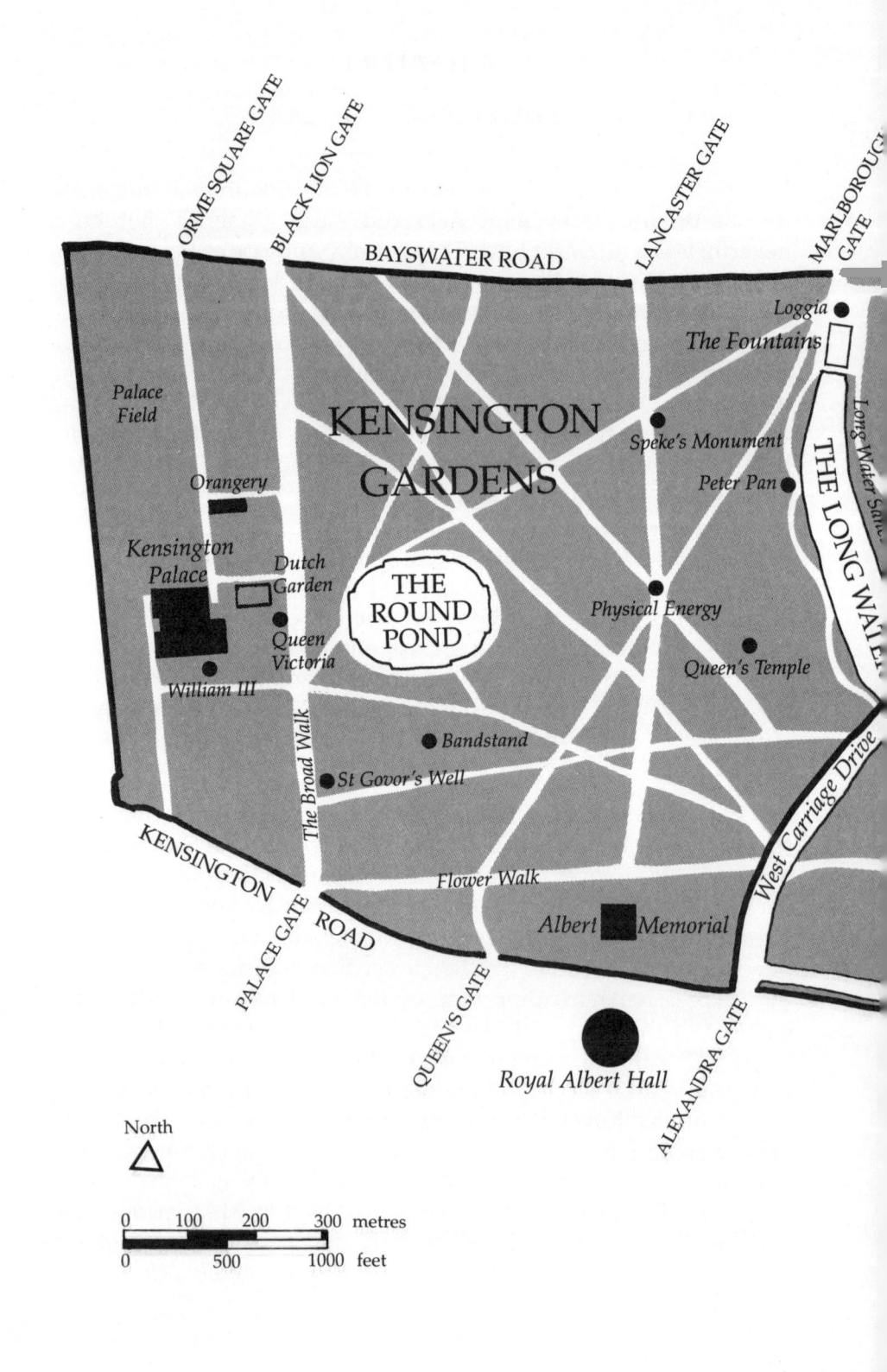

- ORME SQUARE GATE
- BLACK LION GATE
- BAYSWATER ROAD
- LANCASTER GATE
- MARLBOROUGH GATE
- Loggia
- The Fountains
- Palace Field
- Speke's Monument
- Orangery
- Peter Pan
- Kensington Palace
- Dutch Garden
- THE ROUND POND
- Physical Energy
- Long Water
- THE LONG WATER
- Queen Victoria
- William III
- Queen's Temple
- Bandstand
- St Govor's Well
- The Broad Walk
- KENSINGTON ROAD
- PALACE GATE
- Flower Walk
- Albert Memorial
- West Carriage Drive
- QUEEN'S GATE
- ALEXANDRA GATE
- Royal Albert Hall

North

0 100 200 300 metres
0   500   1000 feet

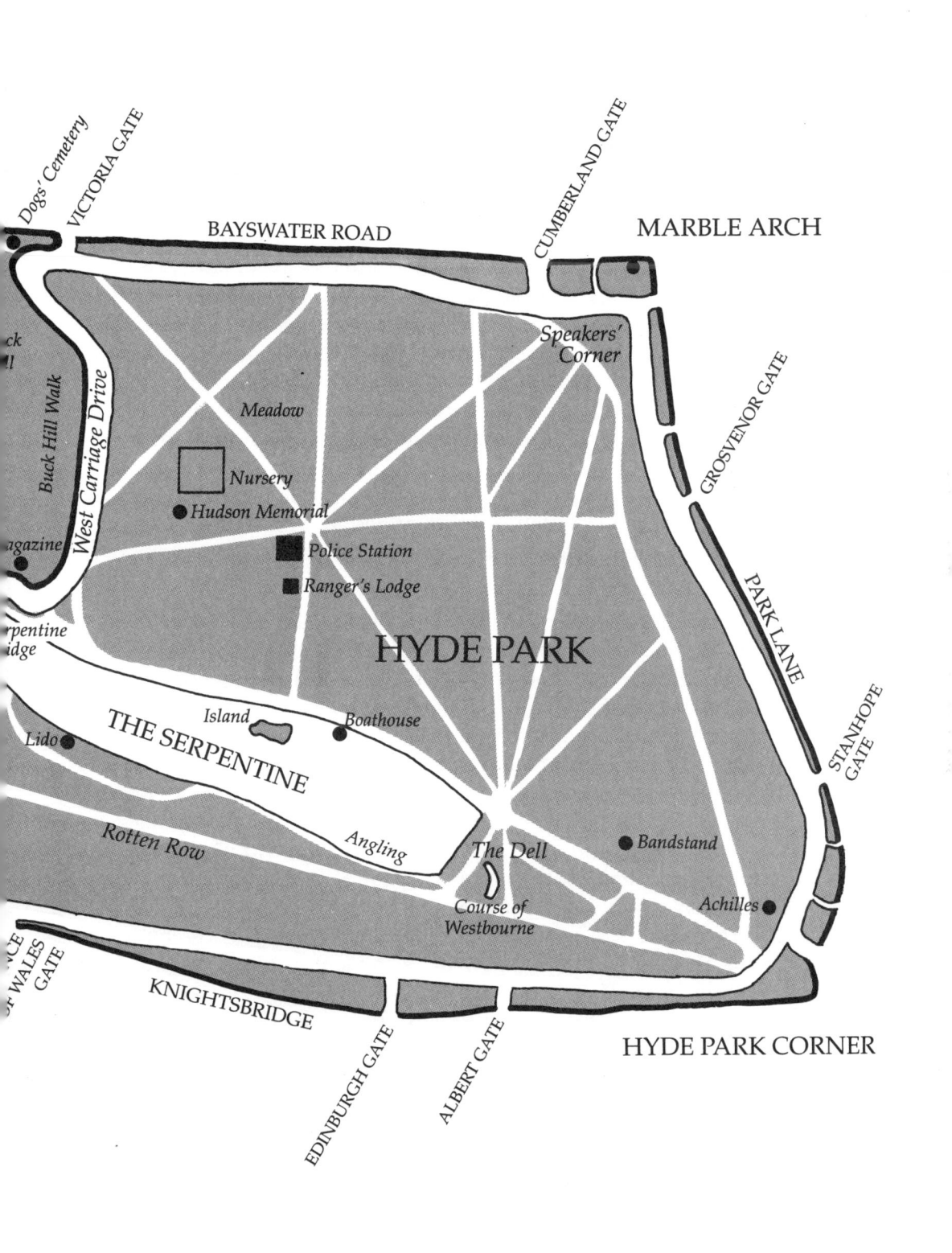

Dogs' Cemetery
VICTORIA GATE
BAYSWATER ROAD
CUMBERLAND GATE
MARBLE ARCH

Buck Hill Walk
West Carriage Drive
Magazine
Serpentine Bridge

Meadow
Nursery
Hudson Memorial
Police Station
Ranger's Lodge

Speakers' Corner
GROSVENOR GATE
PARK LANE
STANHOPE GATE

HYDE PARK

Lido
THE SERPENTINE
Island
Boathouse
Rotten Row
Angling
The Dell
Course of Westbourne

Bandstand
Achilles

OF WALES GATE
KNIGHTSBRIDGE
EDINBURGH GATE
ALBERT GATE
HYDE PARK CORNER

drakes are still a little mortified after their recent degradation, and not so arrogantly sure of themselves as they afterwards become; the old ducks, who have reared the families while their mates hid, and the young ducks of the year are still inclined to be cool towards them. Yet it is the ducks which make the most of the determined efforts to attract. Craning their long necks till they lie almost flush with the water, they swim rapidly to and fro with a fascinating swinging rhythm, describing circles and figures of eight about the rest. The accompaniment of this performance is a chorus of very weak, faint notes, more like the voice of a newly-hatched chick than that of a grown bird. There are two such calls which are particularly common, one plaintive, like the well-known flight-note of a meadow-pipit, the other piping and monosyllabic, like a feeble oystercatcher. So low are these calls, and so delicately uttered, that it is difficult or sometimes impossible to detect the bird in the act of making them. Generally in this sort of ceremony excitement does not run high: the wooing ducks show signs of it, but the drakes, as a rule, are quite ostentatiously uninterested. Sometimes, however, the drakes themselves hold similar performances, and there is a good deal of variety in the details.

Another form of courtship, between a pair of already mated birds, is much more emotional. Swimming face to face they begin deliberately to telescope and extend their tall necks, bobbing up and down with the jerkiness of a mechanical toy, the bill dipping at each bow into the water, which falls off in drops as the bill is raised again. This performance appears to exercise on them a powerful hypnotic effect: sometimes they turn the head slightly to regard each other with one side-glancing eye, sometimes to stare face to face; the *tempo* is gradually accelerated and, instead of bobbing up and down in unison, they often fall out of time and begin jerking alternately, but so fast that the eye grows confused in following their movements. Emotional as it undoubtedly is, this ceremony is performed not with any show of excitement but in a dreamy far-away manner. If it is carried through successfully and without distraction, until the pair are keyed up to the desired pitch, the duck next retires a little way and proceeds to lay her head and neck flat upon the water. She does not give the impression of consciously laying it; rather, she seems gradually to stiffen out and subside in ecstasy as if she had come to the end of a ceremonial dance. On this invitation pairing takes place, the drake seizing with his bill the loose feathers of her nape. She is often entirely submerged. After pairing the drake launches himself off with a curious kind of passive impetus like a craft taking off from the bottom of a water-chute, and in this way he describes a semicircle or more round the hen, his neck stiffly extended, flush with the water, while the hen flaps and performs purification by splashing water all over the wings and body.

These elaborate ceremonies are formal and seem almost invariable, except when they are interrupted or the advances of one bird are refused by the other. Often some person beginning to dispense offerings of food within a reasonable distance will induce the hungrier (or the greedier) of the pair to lose interest in the ceremony and go off to feed. I have seen a duck with an unenthusiastic mate grow very distressed during the bowing performance whenever he showed signs of giving it up and, swimming to his side, bob furiously to rekindle him. She was not very successful. On the other hand, I have seen a duck refuse to take any notice of the drake whatever, in spite of his spirited efforts to dislocate his neck in her presence. It was on the Round Pond; in the end she swam coldly away to show that she was not interested. The drake was so absorbed in his bobbing that for quite a little time he did not notice she had gone. When he awoke to it he turned and followed in her wake, still gallantly keeping up the exercise as he swam, like a convulsive moorhen. A rival drake, approaching from the other side, began similar advances and the duck was now clearly embarrassed. Others of the ducks round about were growing interested and a little excited when she suddenly made up her mind and sprang into the air, the two amorous drakes in hot pursuit and a bunch of four or five inquisitive others following them. They made the circuit once and went off towards the Long Water, leaving me not only the picture of a characteristic incident but a hint of how much right of choice the duck arrogates to herself before the spring frenzy puts her in subjection. This happened on 23rd November. In autumn and winter these performances do not greatly excite any birds which happen to be looking on; but in spring it is very different. Passions run very high and are let loose on the slightest provocation; the jealousy especially is extraordinary; there is a general *noli-me-tangere*[11] atmosphere and if

---

11. ED: i.e. warning against touching.

a drake happens to swim near or in the direction of the mate of another a terrific duel is likely to follow, the rivals whirling round and round in an absolute tornado of wings and splashing water. This jealousy is not so unreasonable as it seems, for they need to use constant vigilance to guard their ducks from one another. They do battle not only for, but over, the body of the unfortunate duck, and the struggles are protracted and indescribable; for both it is a fierce and exhausting experience. But the overcrowded conditions of the semi-domesticated ducks on the Round Pond and in London generally make it impossible to draw the conclusion that their behaviour is the same as in a normal wild state.

Mallard are known sometimes to practise bigamy, but this is a point more difficult to feel sure about in London, where they are so numerous, than elsewhere. I have seen a drake accompanied by two ducks settle near me, one of them keeping close by his side and the other a little apart. This one took wing and he followed, the one who had been at his side remaining alone, without seeming to take any notice of his departure. Possibly this was a case of bigamy; it might be interpreted that the neglected one who sulked apart was designing by suddenly flying off to induce the drake to follow on impulse, as he did, and the other was either displeased in her turn or taken by surprise. But these tempting hypotheses are not always right.

Ducks are more prone to hybridisation in a wild state than almost any other family, and in London there has recently been a most remarkable case of this unnatural mating. A drake **gadwall**, who has now spent six winters on the Serpentine and Round Pond, has for the last four years or so mated with a female mallard and remained all year round in Kensington Gardens.[12] He attends his duck persistently and much more closely than a drake of her own species and is desperately jealous of mallard drakes, though he has had no ill feeling against pochards. If any other bird attempts to pass between him and her he is very distressed about it, 'crunking' and whistling anxiously and showing a warning white patch on the

---

12. ED: This bird was first seen on the Round Pond in the winter of 1920/21 and last seen in the winter of 1932/33 (E. M. Nicholson 1935, Our Parks are Strongholds of Bird Life. *Daily Mail* 7th June). W. G. Teagle has suggested (*in litt.* 1995) that this bird could well have escaped from captivity, perhaps having been bred in St James's Park and been left unpinioned.

wing. Under excitement, and especially anger, I noticed he always lifted the wing so as to give a glimpse of the white speculum, and the more excited he grew the more white showed till the whole of the patch was exposed. I succeeded, by walking rapidly down to the edge at the right moment, in causing them to launch out in alarm on either side of me, but once fairly out he wheeled and rejoined her as fast as he could. About the same time I separated a pair of mallards on the grass in a similar way. The drake took wing and flew over to the Pond; the duck remained where she was. She showed no distress at being parted; I doubt if she noticed it. The quacking or 'crunking' note of the drake gadwall seems to be heard only from January onwards through the spring months. He is decidedly smaller than the mallard drakes, but drives them off with ease.

Mallard often nest, as is fairly well known, in the hollows of the old trees,[13] and the chicks with their downy plumage drop quite harmlessly to the ground, pick themselves up and begin their trek to the water. There they spend a perilous childhood and, in spite of the apparent security of such a haunt as the Round Pond, many of them die young. One family of seven slept last summer in a compact mass on the bare asphalt margin, the old duck a little apart or sometimes standing in the midst of them. Even at midsummer they would doze on like this till nine o'clock in the morning. The reason for this curious bunched formation, rather like the *jugging* of partridges, appeared to be that the ducklings were still too young to sleep on the water and, being flightless, could not roost in trees; by sleeping all in a clump as close as possible to the water's edge they made it impracticable for an enemy to snatch one without giving the alarm, or to cut off their retreat. I once watched them wake themselves up. The first awake shook himself, with a vigour which his sleeping companions feebly resented, and waddled over to join a crowd of ducks and sparrows that a boy was feeding with crumbs. After some time two more awoke and began by their scrambling and fidgeting to rouse the others; all looked miserably

---

13. EMN: In the middle of April I have seen a duck moving precariously along the branch of a tree in the Broad Walk about thirty feet above the ground. She surmounted one very steep almost precipitous slope and finally flew to another tree farther along. Perhaps she was prospecting for a nesting site—she plumped down in one or two major hollows as if to test them—but there seemed no great interest for a duck in either of them.

draggled and unkempt, their feathers being still at a transition stage from down and very easily ruffled. Most of them on first waking stretched their wings, which were grotesquely small and had the appearance of being set unnaturally near the neck, since the pinions to give them balance were not yet grown. Patches of bare flesh showed at their bases; they were like angels' wings in their incredible position, but not in any other respect. Simultaneously with the wings they stretched out their full length, the right foot being extended to its utmost behind them. Their first actions on rising were varied. Some rushed over to the scatterer of crumbs for breakfast, others plunged briskly into the water and struck out; one took three long drinks, the first while standing on the bank and the others as he swam.

Once, seeing the flock of sheep working along the margin towards the duck's family, we[14] shepherded some of them to pass near her to see what she would do. One halted and put its head down, staring stupidly in her face. She raised herself in defence with half-open bill, reminding us both very strongly of a viper on guard. Then, when a second sheep arrived, she slipped off with her brood into the water. I noticed another piece of duck psychology at the same spot a few days before. Here at the east side was a good deal of flotsam and jetsam drifted by the prevailing winds, and among it a dead duckling, perhaps a fortnight old, which bore no visible trace of hurt. The family to which it had clearly belonged came waddling along the margin—the ducklings were the same age exactly, and by this time most other families were much further advanced, since it was now July. When they came to the corpse we watched them narrowly, but they showed not the slightest sign of recognition. One of the brood actually stumbled over it and, after several unsuccessful efforts to scramble past, went round it and on without turning a hair. So far as he was concerned the obstruction might as well have been a log.

During this flightless stage of their existence, the 'flappers' are most amusing to watch. They exhibit in working order habits which the adults have forgotten. One of these is the capacity for diving. The mallard, of course, is a surface-feeding duck, belonging to the same main division of the family as the shoveler, wigeon and

---

14. EMN: i.e. my brother and I.

teal, while in the other are included all the diving ducks, the pochard, mergansers, tufted duck and goldeneye. The distinction, however, is not properly between ducks that dive and ducks incapable of diving, but between those which dive habitually for their food and those in which the faculty of diving, though present, is dormant and not in regular use. Now in London this dormant faculty is often revived by the mallards, and then both the ducks and the drakes dive freely and well. One may see nothing of it for weeks and then come down one morning to the Round Pond and find half a dozen or more plunging constantly and remaining submerged for appreciable intervals. One duck I noticed was a particular expert: her dives must have lasted on average more than half a minute, and they were very frequent. As she swam she would suddenly duck her head and, at the same time impelling herself by powerful strokes of the feet, dive for a distance of three or four yards. In the clear shallow water her form could be followed all the time, though sometimes only faintly, as she moved two feet or a little less below the surface. Generally she rose in a natural posture, but sometimes the stern, which was always during the submersion at a higher level than the downward-striving head, would come to the surface first. Once she made a complete circle, passing beneath a drake who was floating on the surface. She dived to good effect, for twice or three times she caught some small silvery wriggling fish, not more than an inch long, which she brought to the margin to be eaten. They were extremely lively creatures: one of them reappeared half a minute after being laboriously gulped down and was only by a tail's breadth prevented from escaping. The dives could neither be described as amateurish (though false starts were often noticed) or superficial, for they went as deep as the shallow water permitted, and there was no trace of surface ripple. By taking down some maize and throwing it into fairly shallow water we succeeded in getting adult mallards to dive freely for it and to display a surprising ability in exploring the bottom. But they preferred surface food and tried if possible to intercept it before it sank. A few of the drakes were very incompetent divers, delaying too long between putting their heads down and continuing the impetus with their feet, so that by the time they struck out they were standing on their heads in the water with their feet splashing vainly into the air, making a needless commotion before they finally submerged.

The 'flappers' have this faculty far better developed than the older birds. They dive constantly, even when they are only two or three weeks old, reaching a depth of two feet or more in the shallow basins at the Fountains in Kensington Gardens,[15] where their movements under water can be followed perfectly. Compared with the placid sedentary duck they are marvels of energy, racing about and snapping at anything they see, from flying insects and flowers of the rush to weeds at the bottom of the water. Their action, at full speed, is not unlike the swimming stroke known as the crawl, and they seem to attain by it a much greater pace than when they are adults and use more conventional methods.

In only one comparatively rare performance do the old birds show the same exuberance as these 'flappers', and that is an extraordinary exhibition the significance of which, if it has any, has not so far been determined.[16] It is a kind of mad horseplay in which both ducks and drakes take part, but if it is in any way connected with courtship nothing happens to give a clue to the fact. Assembled in large scattered parties first one bird and then another will spring up forcibly—feebly into the air, struggling along a few yards and dropping heavily with a splash, as if it had been attacked by cramp. A sudden uproar begins, some standing up and flapping violently upon the water with a loud galloping noise, others making the same strange crippled flight and, on dropping, plunging completely under again and again, screening themselves at each rise to the surface by a mighty splashing of water over their backs. Sometimes drakes pursue ducks, sometimes ducks drakes, sometimes single birds flounder by themselves. It is sheer madness and there is no trace of method in it. Other ducks have this habit, and even on the Serpentine I have seen it done on a smaller scale by the tufted duck.

It is possible to trace the stages in the increasing shyness of the mallard as you leave London behind. On the Round Pond they feed from the hand; on the Pen Ponds at Richmond they swim quite fearlessly near the bank, but they are much less trustful; on the Heron Pond in Bushy Park they have taken wing and flown to the

---

15. ED: The Fountains, at northern end of the Long Water, are now known as the Italian Gardens.

16. ED: For a possible explanation, see T. Lebret 1948, The "diving-play" of surface-feeding duck. *Brit. Birds* 41: 247.

other end when I walked towards them; on Staines Reservoirs they seem not to be afraid of passers-by but to rise, even at a fair distance, if you stop and look at them. Farther out, especially where the wildfowlers work, it is a question of seeing them before they see you. Even in Kensington Gardens their confidence varies with the hour. Towards nightfall they become suddenly suspicious: if food is thrown they spring up in alarm or push rapidly out into open water, quacking uneasily. In the early morning they are equally distrustful and, tame as they are in the middle hours of the day, they will often take fright at something on the bank, especially a large dog. Then, at a quacking signal of suspicion the whole company launches hastily out towards the centre of the Pond, swimming in line abreast. In autumn, in their fresh plumage, and when some numbers are present—I have seen a navy of forty-one retreat in this way with perfect order—the incident is almost spectacular, especially since in a wild state it could never be seen, for then if the observer had succeeded in coming so near all would spring up in confusion.

On the morning of 14th November 1925, when the fog was so dense that the extreme limit of visibility in Kensington Gardens was not more than fifty yards, and the cold so severe that even at this early date the greater part of the Round Pond was frozen hard, I watched with unsympathetic amusement the struggles of the dazed mallards against the new-created element. Arriving at the margin of the ice a duck found the way barred for swimmers and so attempted to land. Ducks, of course, have their feet set so far back that—unless it shelves—they take to land in two motions, first heaving up the breast and then getting enough purchase with the feet to enable them to find their terrestrial balance and walk. But on the ice that was not feasible. Whenever she levered her breast onto the ice it gave slowly and snapped with a metallic cracking sound just as she was on the point of hoisting herself up. This happened time after time, till she had ploughed through two or three yards to a part where the ice was thicker, and there with a last effort she succeeded. She marched along with a timid comical waddle, slipping at almost every step, for the broad webbed soles gave no purchase. The objective was a large crust of bread lying on the ice. Scaring away by her approach the sparrows which had been sharing it among themselves she gave it a peck, but it was hard and her broad bill could make nothing of it. The effect of her clumsy violence, and of the blundering approach of another duck more awkward than herself, was to smash the frail ice on which she was standing, leaving crust and duck afloat on a little pool of water. With a fresh determined onslaught she succeeded in ramming the crust right underneath the ice. Next followed an attempt to recover it, exasperating for her but farcical to the observer. If only she had had the intelligence to break the ice beyond the crust, which she could easily have done, she would have got it back at once. But she persisted first in trying to clamber on top of the ice above, which invariably gave way and forced the crust farther under, or in craning her neck beneath in a clumsy attempt to pull it out, which accomplished nothing beyond pushing it farther away. At last she left it and swam back to the others; she seemed at first to act like a decoy, trying to persuade them to follow her back up the channel to the place where the crust had been lost, but only her mate came and nothing more was done.

On the Serpentine I have seen a mallard alight safely on the ice, but as a rule they dislike it, and they abandon the Round Pond as soon as it is frozen over. They are subject to the weather as all birds are, and at the Round Pond the side where they are to be found seems to be decided by the direction and strength of the wind. If it is light they prefer, in the absence of stronger attractions, to browse among the flotsam and jetsam on the windward side, but if it is strong enough to inconvenience them they keep to the lee. The tempests which they meet on the Round Pond are not to be despised: I have seen rollers a full eight inches high breaking on the eastern shore of the Pond and sweeping over as far as the seats on the edge of the grass.

They fly from place to place freely, especially at dusk, when they pass so low down the glade past *Physical Energy*[17] that to stand in the fairway is almost dangerous. Their method of descending is fascinating to watch, and it throws light on the use of the tail as a brake. The weight is in heavy relation to the wing-surface and the bird hurtles through the air with terrific momentum. At his

---

17. ED: *Physical Energy* is a huge equestrian statue, erected in June 1904, by George Frederic Watts of Cecil Rhodes. The statue is situated between the Round Pond and the Long Water in Kensington Gardens; it is a bronze cast of the original which forms the centre of the Cecil Rhodes Memorial on Table Mountain at Cape Town. A photograph of the statue appears on page 27 of *London's Royal Parks: An Appreciation* by Richard Church (HMSO 1993).

ordinary flying speed an impact with the water would be severe. To prevent this he descends with the whole body acting as a tail, the head and neck alone remaining horizontal, and so checks his impetus. If his tail bore the same proportion to his length as a great tit's it would need to be ten inches long, which for a water-bird would be very inconvenient. Most diving birds have short tails and long necks, which like the mallard's serve as a substitute in flight. The amphibious gulls have normal tails but swim with them very much raised. Clearly there is a rough ratio between the size and weight of a bird and the tail-surface required to draw up in a normal way. All the larger birds which often perch on trees have tails which are either very long or, more usually, with a great reserve of surface ready to be fanned out. The woodpigeon and other doves are conspicuous examples, but hawks, crows, cuckoos, thrushes, blackbirds, shrikes and many others show the same thing. The contrast between gamebirds that perch on trees—such as capercaillie, pheasant and black grouse—and those living on the ground—like the partridges, quail, red grouse and ptarmigan—is especially striking.

Soon after the middle of May the drakes become distinctly shabby and show signs of the approaching moult. I have seen one which had lost most of the chocolate breast-feathers by the 23rd. By July they are a disgusting sight, in all stages of living decomposition, as repulsive as leprosy. The very brightness of the normal plumage makes its decay more hideous; one sees, for example, a drake with half his head still bottle-green and the other half naked like a vulture's. At the outset of September the more advanced birds have reassumed spring plumage, although the curled tail-feathers do not generally appear till about the middle of the month. The young drakes of the year become now plainly distinguishable from the ducks, but even by the middle of November some have not yet gained their adult dress. At this stage the transition plumages are extraordinary: a bird will be found with the mature green head and white collar, a mottled but distinctly chocolate breast and, on the flanks, mixed with the developing silver-grey feathers, large patches of coarse brown duck's plumage, quite unmodified. Even when the moult is complete there are many striking varieties. One drake was devoid of white markings, the bottle-green head joining the chocolate breast with a curious effect; another had a triple depth of white collar, white on the nape, and a patch on the tip of the wings. A glaucous bill is sometimes seen instead of the orange-yellow, and a drake on the Round Pond possessed neither a white collar nor a chocolate patch on the breast, the whole plumage up to the bottle-green head being the same quiet grey as the flanks and most of the mantle. Ducks also vary in shade and sometimes have white on the wing. It would be possible to make a tedious list of these varieties, for they cannot be called scarce, but the subject did not greatly interest me.

In inner London[18] at any rate the **pochard** is exclusively a winter visitor. I have not seen it before October on the Round Pond. Shy on first arrival it rapidly gains confidence, becoming in a few days almost as tame as the mallards, though when people are present it will not come on shore. It does not hesitate to jostle the mallards and snatch their food, though they are larger and have more powerful bills, but it suffers from the attendance of piratical gulls as much as the tufted duck if not more. It swims low, with the neck almost awash.

Considering that these pochards are true wild birds, and perhaps belong to almost uninhabited countries, they learn confidence extremely fast and put up with a great deal. A few days after they reached the Round Pond an elderly idiot introduced a craft which looked on a side view like a model boat with a sail, but was in fact only a painted board, a kind of silhouette, propped up on the water by an arrangement of outrigged corks, which he kept like a kite on the end of a string, so that it could be half sailed, half dragged, across the Pond by the resistance of the wind. At intervals it plunged completely beneath the surface with irresistibly comic deliberation, in a curve like a sea monster diving. The string by which this curious craft was towed proved a great plague to the mallards and pochards, for as the idiot moved along the bank the string, keeping pace under water, kept threatening to entangle their feet. I was astonished that the pochards tolerated it, so soon after their acclimatisation, but they did.

---

18. ED: The phrase 'inner London' is used in this work, except in the Appendix, in a loose sense and should not be confused with the Inner London recording area—a rectangular area, five miles north to south, eight miles east to west, centred on the site of the old Charing Cross on the south side of Trafalgar Square—which was not defined until 1928 (see M. J. Earp 1991, The Origins of the Inner London Recording Area. *Lond. Bird Rep.* 54: 138–140).

The twinkling red eyes of the drakes grow redder and more twinkling under stress of excitement. By day they bask lazily on the water, sometimes preening, sometimes asleep, although they often come to shore for food, but they are galvanised into new creatures by the approach of night. On the Round Pond at dusk I have seen them diving continually in quick succession, close up to the margin where the food was sunk, so that often the water boiled like a spring with their submerged activity, and a half a dozen of their grey mantles could be made out at a time pushing along the bottom.

I have noticed courtship in early March; it bears some likeness to the mallard's, but is less elaborate. There were vestiges of the bobbing performance, and such traits as holding the neck stiffly erect or swimming fast with it resting on the surface of the water. The one really distinctive action was the characteristic toss of the head. The language which accompanied these demonstrations was like the mallards' in being a series of undertone-notes, and gave the same impression of coming from chicks or little passerines, not adult ducks. One of these notes sounded curiously like the redpoll's mellifluous 'Honree'; another was a rather prolonged agonising whisper, and there was a sound resembling the faint quacking of a mallard, beside the true call, an almost crow-like 'Karr'.

The **tufted duck**'s note is very like this last, and at Littleton Reservoir in June I have heard it very frequently uttered—a distressed grunting croak. But the tufted duck as a rule is both silent and undemonstrative and, although I have watched them at all seasons, I have never detected any trace of courtship ceremony. Mr Boase, who has looked for it in Perthshire, has been rewarded by the most trifling and occasional glimpses of bobbing, head-tossing, and calling in undertones; he thinks[19] most of the activity must take place at night. I have seen many signs of animation in the birds which remain all summer on the Serpentine, but they were connected with jealousy rather than courtship. On 22nd May (1925) I saw a drake come off the island and swim over to a raft where his mate was sitting among mallards. He scrambled up and made a hostile demonstration, with the neck extended and bill open like a goose, then flew over their heads into the water. The next day I

---

19. EMN: H. Boase 1926, Notes on the courtship of the tufted duck and its distribution in Scotland. *Brit. Birds* 19: 226–230.

found two pairs about the island, very jealous and irritable, one couple being in possession of the only available nesting-site and unwilling to share it. This first pair considered itself to have sole proprietary rights over the island, and less strictly over the raft; the others were tolerated on the water, provided they kept their distance. The result was incessant warfare, very amusing to watch. One drake would swim towards the other, his eye becoming, it seemed, quite observably jealous and flashing hate as he accelerated, till he flapped along the surface at full pelt. The other, without waiting for the impact, would often at the last moment dive neatly under him, like a man side-stepping from a charging rhinoceros and, coming up behind, take wing for safety before the first had time to turn. The ducks were not regarded as non-combatants, and one of the drakes, fastening on to the tail of the other's mate, pinched her mercilessly for an appreciable time. Once the attacking pair scrambled on shore unobserved, while the others were dozing on a raft. Suddenly the possessing drake woke up and saw them disappearing among his own private bushes on his own private island: he took to the water instantly, swam across, landed and waddled over, first sending the enemy duck scuttling back in terror to the water and then returning to retrieve the drake, which he succeeded in doing. But instead of tamely retiring the repulsed drake flew straight over and made a base attack on the duck of the pair in possession. She fled across the water, getting badly knocked about; her mate came promptly to her assistance just as she was forced to take wing, and all three of them flew once round the island before the assailant desisted. These hostilities continued until June.

On 14th July I saw a duck on the Long Water with her family only a few days hatched. By the end of August the five which survived had come to weeks of discretion, and were compact rounded little ducks of an almost uniform grey-brown, feathered but still distinctly downy. It was one of these ducklings which I first noticed preening its belly, a common and amusing performance among tufted ducks, which looks to the casual observer as if the bird was chasing its tail. In this delicate operation it is necessary for the bird to careen until it floats bottom side up with one helpless webbed foot waving feebly in the air while the head is manoeuvred by much craning and twisting of the neck to reach the desired feathers. The other foot, still under water, is evidently used desperately as a paddle in an attempt to keep the balance, but working alone and at unaccustomed angle it makes the bird spin round faster and faster on its own axis till it becomes too giddy to continue. After a short rest another attempt is made, till the feathers are considered satisfactory.

On the London park lakes, where the water is moderately clear, it is often possible to watch from beginning to end the diving of the tufted duck. Taking a sudden powerful plunge the bird propels itself swiftly but at an easy grade—by no means steeply—down to the mud and gravel bottom. The neck is fully extended as in flight and the wings lightly folded to the sides. After reaching the bottom it progresses in a curious manner, the disc-like bluish paddles treading water backwards at full pelt to keep its buoyant form from shooting to the surface, and the tail in spite of these efforts much higher than the head and bill, which busily prods and pokes for scraps of food among the mud and stones. The feet, it must be explained, operate not in the usual position but by the side of, if not actually above, the stern: they have almost the flexibility of a ball-and-socket. The feathers of the stern release at intervals a series of bubbles, which mark the bird's progress when its form is temporarily invisible. It travels fast under water and turns freely, without the least awkwardness. I have pretty well satisfied myself that it always keeps an eye on what is happening in the over-world, and whenever there is any sudden movement on shore to excite suspicion it will cease paddling and shoot obliquely up to the surface. I have often seen one dive under such a crowd of swans and mallards and other waterfowl that it had difficulty in finding a place to come up. The lack of curiosity shown by people generally towards this fascinating operation carried on under their noses amazes me: even those who bring food and seem to take some interest in the birds rarely make the least effort to follow their movements under water, though they often dive within a yard or two of the bank. Yet any sort of a machine in a shop-window has an audience all day long.

The appearance on the wing is of a small sooty-black duck with an impure white hind margin to the wings. It appears much faster than either the mallard or the pochard (though allowance must be made for the smaller size, which would tend to produce that illusion) and is also very erratic, swerving almost as frequently and abruptly as snipe; in taking a sudden turn the feet are sometimes used. I estimated the speed of one flying repeatedly backwards and forwards over the Round Pond at not less than 70 m.p.h. and, though speeds are often deceptive to judge, this bird gave an exceptional opportunity.[20] On March passage at the same place I have seen about twenty rise in company with half a dozen or so pochards and fly backwards and forwards in formation. Many were already paired (3rd March); but I have seen a drake and duck get up and fly round together on 25th January, as if they were mated, and three days later a pair by the Serpentine island gave very much the same kind of mad horseplay as I have described in the mallard, continually flapping along and diving again, so that both were never in sight together. The mallards and other tufted ducks took no part.

They go into eclipse plumage about July and seem not to emerge from it till considerably later than the mallard; in 1925 I saw no more in full dress till 18th October. During this time the white panels on the flanks turn to a light biscuit shade and the long plumes of the head are lost. When these appear again they show in the sun a green as well as a purple gloss.

In winter, and especially on passage, there is now hardly a water in the metropolitan area where they are not found; they even haunt the little South End Pond on Hampstead Heath, which is bordered by houses. The numbers on the reservoirs are often immense: on Staines alone I have seen at least two thousand together.[21]

Other ducks visit the neighbourhood of London more or less regularly, some, like the wigeon, penetrating even to the Serpentine in hard weather. The waterfowl which though artificially introduced live an almost equally wild life in the parks hardly come within the scope of the observer, although he is always freshly astonished at the spectacle of a pair of droning Egyptian geese in full flight round the Serpentine island, or a large and a small Canada goose on the wing together over the Long Water. I have seen the big Canada and a Chinese goose swim face to face

---

20. ED: According to the article on Speeds of Flight in *A Dictionary of Birds* (B. Campbell & E. Lack 1985), the fastest reliably recorded speed of steady flight by wild birds is well below this, at some 47 m.p.h.

21. ED: See also E. Gillham 1986, *Tufted Ducks in a Royal Park* (St James's).

and begin bowing elaborately, like bobbing mallards except that the neck was bowed instead of telescoped; they were so close that they seemed to get in one another's way. Pairing afterwards took place, the Canada being the gander, and this union resulted in a hybrid brood of goslings. Shortly afterwards the ceremony was repeated by a pair of Chinese geese not far away. These, unlike the Canada geese, are pinioned and cannot fly.

CHAPTER II

THE LONDON GULLS

As a family the gulls are among the most interesting British birds, for in their status, and consequently their whole habit of life, they are among the least stable. We still call them sea-gulls—at least we do in London—although they are, in fact, tied to the neighbourhood of land by the very nature of their diet. They are, it is supposed, lost tribes of the plover race which developed by specialisation a more aerial and aquatic constitution, until they gained the powers, which plovers have not, of soaring without effort and of swimming naturally and easily with their webbed feet, and of feeding indiscriminately on all edible things. It is also supposed by some, though with far less reason, that a reverse tendency has already set in, and that future evolution will draw them back to the land. To discuss these theoretical tendencies is outside the province of a field-observer, who sets himself simply to put down and comment upon what he has seen.[22]

Six species of gull breed in the British Isles, and all of them except the kittiwake are in a sense London birds. One, however,—the British **lesser blackback**—is never a permanent or a significant visitor and may fairly be given short shrift here: it occurs on migration, passing overland as travelling gulls often do, and I personally have identified it only once in the metropolitan area—an immature unlucky wanderer with a broken leg hanging stiffly down, who passed over the Long Water in Kensington Gardens on the morning of 15th September 1925.

There is a good reason for the comparative rarity of the lesser blackback in London, for gulls frequent the place only in winter, and this alone, of the British species of its family, retires at that season from the British seas.[23] A week later, on the Kent coast, I watched them come swinging along one after another just above the waves, rounding the corner of England till off Hythe they were heading due west, when the bend of Dymchurch Wall turned their

---

22. ED: For a nearly contemporary alternative account, see S. Gordon 1935, *Sea-gulls in London*.

23. ED: This is no longer true—see page 169.

heads south again and they followed the east side of Dungeness out to the extreme point where, apparently, they began the passage of the Straits. The majority seem to have their winter quarters off the coasts of Portugal and Cadiz Bay; some go even farther south.

The **greater blackback** is proudly claimed as a regular winter visitor to the Thames, so high as the reach off the Embankment between Waterloo Bridge and Blackfriars, in the heart of the County of London. Grand and conspicuous as it usually is, it contrives in my experience to be unaccustomedly elusive in the metropolis, and I have set eyes on it only once, one day late in January, when a majestic adult took wing in midstream and flapped stolidly away with characteristically drooping wings.[24] If its behaviour in London differs appreciably from its behaviour elsewhere, which seems unlikely from the small numbers that visit us and the very brief duration of their stay, I have no notes on the subject and it must be passed over.

The **herring gull**, in many parts the dominant species, is not so in the metropolitan area, but it is present on the Thames from the beginning of November until the end of March in sufficient numbers to rank as a true inhabitant. It nests on the grassy slopes rather than the actual precipices of cliffs, as at Lyme Regis landslip in south Devon where I have overtaken and caught on the beach a flightless young bird which must have scrambled down from the nest unaided. At Lyme Regis they settle freely on the house-tops, uttering their dry cynical laugh, and on the sea-front allow themselves to be approached within a few feet. In London, by contrast, the herring gull is not a trustful bird, and on the river or artificial waters he usually keeps well out in the midst, at a safe distance from the people. He belongs in fact to that curious group of birds which are actually less approachable in London than elsewhere; in another chapter the same thing is noted in the case of the robin and the chaffinch.[25]

---

24. ED: In *Birds and Men* (1951) EMN wrote that 'During the last few years there has been a conspicuous increase in the visits to central London of the great blackback, which was recorded there only four times in the whole of the nineteenth century, became a rare but annual visitant just over twenty years ago, and is now to be seen daily. On 29 January 1950, E. R. Parrinder, C. B. Ashby and I counted 235 on the Thames between Barnes and Woolwich.'

25. ED: See page 102.

The London Gulls

The herring gulls which come up as far as London show a striking preponderance of adults. On the coast, where they are in most parts the commonest species, half or more than half the total strength is often found to consist of immature specimens in the brown and, since it is not till the fourth year of their existence that they assume the pearly-grey mantle of the adult, this proportion is hardly surprising. But in London I do not remember ever having observed an immature bird except on the river, and even there the adults are as a rule conspicuously dominant. Their arrival in the parks is more or less dependent on sharp weather; when the Round Pond was frozen hard I have heard the characteristic loud laugh and seen three together standing in the centre. On Lord Mayor's Day last year (9th November 1925), after a night when eight degrees of frost had been recorded, there were at least four herring gulls already present on the Serpentine. This, I believe, is an early date, for they do not appear to reach London before November.

On the coast, where the sea gives him a true proportion, the herring gull is not a particularly large bird; but stand on Blackfriars Bridge and see him wheeling in midstream above the swarm of lesser species—here among the little birds and over the little murky river he appears as grand as an eagle. I have seen one standing on the Pelican Ponds in St James's Park lake which fascinated me for many minutes by his illusion of great size; he was alone, but if the pelicans themselves had been there they could not have looked more stately. Such things are relative and must suffer continual adjustment; in London we take our standard from the black-headed species and by it even the woodpigeon counts as a really large bird. This, the **black-headed**, is the typical gull of London, and he outnumbers all the other species by at least fifty to one. The main arrival in autumn is by flocks; it took place in 1925 in the first week of October,[26] and on the 7th I witnessed what may have been the first appearance at the Round Pond. They came over from the east flying high and easily against a sombre sky, late on the wintry afternoon of that day and, seeing the water beneath them, began to descend, slowly at first but in the end all plunging down together. From this time I began to take a great interest in the black-headed

---

26. EMN: It is by no means regular: numbers were noticed many weeks earlier the following autumn.

gulls of London, and to watch the species at the Round Pond and elsewhere whenever I could create an opportunity.

For the week after they had first been seen they were very little in evidence; on the 13th my brother found only three on the Round Pond and none on the Long Water, though I was told they were already strong on the Serpentine. On the 14th, however, on reaching the Round Pond about twenty past ten, the weather being fine but sharp, we saw numbers beating up and down above the water. When we began to feed the sparrows and ducks they quickly converged on us, flying up and down and pouncing on any pieces which fell in the water, but they would not even attempt to catch one before it had fallen, however temptingly it was tossed up in the air. This unwillingness, or inability, to catch anything on their first arrival was remarkable, for later on they would feed from the hand. When our supply was exhausted an old woman, rather bent and badly dressed, came down and began to fling into the water some stuff which looked like grain and which, instead of floating, sank at once to the shallow bottom. The gulls immediately began to plunge after it, rather in the manner of terns, for they checked their flight and hovered a moment before dashing themselves down with wide-spread wings, the tips of which, and sometimes of the tail also, often remained visible some distance apart above the surface.

But the dives of a few birds were complete and they absolutely vanished under the water for an appreciable interval. All seemed to take part, sometimes four or five of them diving simultaneously from any height between one and six feet, or perhaps higher. Nothing but that impetus carried them under water; it was not a true dive and if the bottom had not been shallow they could never have reached it. In a few minutes when the meagre supply gave out they ceased to plunge and swam idly on the water, the woman having departed before they began, but occasionally in the centre of the Pond I saw one plunge so that only the pointed tips of the white wings remained in sight. In the afternoon, the weather having become dull and overcast, I obtained some maize and went down again to the Round Pond to experiment, but only three gulls remained present and they, though they beat up and down, refused to dive for it. I believed at the time that this must be because without the sun it could not be distinguished on the bottom. The next day, 15th October, we threw in quantities of maize, which the gulls took, though not very eagerly. They could not be persuaded to dive for it. Later, however, they showed more enthusiasm and sometimes plunged freely to retrieve it. At first I occasionally saw a bird hold the grain in its bill a few moments and then let it drop, rejected, but in every case I watched on the 22nd the maize was readily gulped down.

Now if the black-headed gull will eat such a tough and difficult substance as maize, the largest and hardest of all grains, it is obviously a serious potential enemy to corn-growers, though in practice it seems at the present time to be mostly beneficial. It was this point that I had been so anxious to settle, for if a bird is capable of feeding on grain it requires only sufficient pressure from competitors or a scarcity of alternative foods to establish in it a habit which might cost the country a tremendous sum afterwards. The black-headed gull has increased extraordinarily fast during the last fifty years; it would be well for his good name that he should not go on increasing much longer, for then a wider diet would become inevitable.

But there was another subject of the experiments: the light they were to throw on the diving habit in gulls. Just at this period Mr Frohawk and others were carrying on in *The Field* an interesting discussion about this same point,[27] from which it appeared that either the diving habit had been curiously overlooked until recently or that it was now becoming more frequent. Few instances of a total immersion were brought forward, but on the Round Pond I could count six cases of this in only about two minutes, and in addition I had found a method of persuading them to do it whenever I liked. Bacon-rinds, it turned out, were a much more tempting bait than maize. In fact it was necessary to act with great cunning in order to get them to the bottom, for the gulls were possessed by such an unreasonable gluttony for rinds that they would freely and without hesitation glide past and snatch them adroitly from the outstretched hand, within a fortnight of their arrival. A generous bait of bread in the same position left them absolutely unmoved and until it was tossed up they refused to accept it.

They had regained the lost art of catching food on the wing within a few days of their first appearance. Their appetites varied

---

27. ED: See J. R. Harding 1925, Diving of Gulls. *The Field* 146: 697; A. H. Macpherson 1925, Diving of Gulls. *The Field* 146: 771.

and on some days they were shy and stand-offish. Generally the first suspicion of food would bring them rushing across the Pond, and it was almost impossible for the rinds to have time to sink before they were seized and gobbled up by the vanguard, but on these other occasions they would stand sullenly by and watch it all go to the bottom. Then, as soon as we moved off they would start plunging for it, till the whole of the sunken treasure had been salved. All the first month of their stay diving was freely practised and by mid November some of the birds seemed to have increased in skill and daring, for we induced them to plunge for bacon-rinds to a depth of about two feet, which they achieved by hovering and dashing themselves down from as many as five or six feet above the surface. They could be seen underneath when the water was clear; in the most perfect submersions the wings were as nearly closed as a plunging gannet's, and the impetus flung up a proportionate jet of spray.

I was reminded of a fine day three months earlier when I had lain on the most westerly cape of Europe[28] watching gannets fishing out in the Atlantic. The great bird would cruise along about fifty or seventy feet above the waves, his buff head and bluish bill so sharply down-turned to discover his prey that it looked like a gigantic white woodpecker's and his long wings beating with an almost ostentatious ease, the stiff narrow tail preserving the balance. He would fly round in this way, sometimes for several minutes, sometimes for only a few seconds, before suddenly, and with so little warning that it was all but impossible to put up the glasses in time, he would tilt into a rapid slanting plunge, folding his wings sharply backwards to lessen the resistance and striking the sea with such a smack that he instantly vanished under a tall jet of spray. Sometimes, in spite of the distance or the sound of the spray, the sound of the impact could be heard. He would rise again not far from where he had vanished and in the expected direction, preen himself a moment and sometimes shake out his wings as he rested on the water and then take flight, lifting easily (without any threshing of the waves) into the wind, and resume his cruising. Although the speed of the plunge and the final impact were terrific, and seemed to me one of the few common instances of a comparatively large creature acting with as much violence in

---

28.   ED: Slea Head, Co. Kerry.

proportion as a wasp or a grasshopper, it was all achieved without a single stroke of the wings: at the sight of his prey he simply released the plummet of his body, which he had been holding poised in readiness, and nothing else was needed.

The perfection of that scene, lying there on the warm ruddy cliffs with nothing to the west but the flecked ocean and an archipelago of fantastic islets, with the Skelligs like a ship in full sail on the horizon and these phlegmatic fishers, dazzlingly white in the full sun, cruising and plunging as far as they could be distinguished, was far beyond the rivalry of anything in London. Yet on a sunny day of autumn or early spring, coming down to find the Round Pond mirrored blue under the sky and the gulls, like little gannets of this little sea, dashing themselves valiantly down, there is a quality in it so fine that it comes almost with a physical shock in such a setting.

But there is a fundamental difference between the plunge of the gannet and the plunge of the black-headed gull. The gannet, as I have tried to show, cruises along and gains his impetus simply by releasing his full weight from a height, for he is built like a cormorant. The build of the small gull is the very opposite: it is nearly as light as a swallow in proportion. Therefore it has to be impelled, and even then it must come to the surface immediately the impetus is lost. Partly for this reason, and partly because they generally plunge with spread wings, the submersions of gulls are normally incomplete. But some of those at the Round Pond became expert enough to leave no doubt that if the habit should become fixed and concentrated they would develop their powers of retrieving food in shallow waters to rival the down-craning mallard. I have described already how the more expert learnt to fold their wings: even when they aimed deep these were generally successful. Like gannets they would sometimes rise and float on the surface instead of taking wing again. Another ingenious method gave fresh evidence of their efficiency: coming down when there was a high wind I found them taking advantage of it to plunge obliquely from a foot or less above the water, so that the force of it striking them at the proper angle lent as much impetus as if they had plunged from five or six feet in a calm. But on this same morning others were still wasting their energy on the ordinary method.

My experiments at causing gulls to dive brought me quite by accident in touch with a habit which I have never seen mentioned, and which has obviously an important bearing upon the theory of territory. Going along the margin to sink some diving-bait at a fresh point I found my intentions opposed by an adult black-headed gull, recognisable by the exceptionally conspicuous vestiges of his[29] hood, who had seized for himself a piece of territory about twenty yards in diameter centred on the pipe at the north-east corner of the Pond. In this little kingdom he floated on the water with his head and neck drawn up into a crouching posture and uttered a harsh and guttural warning 'Ouarrh-h-h!', prolonged with a slightly vibrant quality, when any trespasser of his own kind approached. Those flying over were generally allowed to pass, but any which alighted in the territory were immediately assaulted so brutally that they always retired. This inhospitable spirit did not suit me at all, for I had already unknowingly sunk the last of the precious bacon-rinds within the prohibited area, and the warden not only showed no disposition to dive for them himself but at once drove off any other gull which attempted to hover or betrayed the faintest interest in them. In order to overwhelm him I threw in some bread and attracted a crowd of gulls but, powerless as he was to keep at bay a gathering of hungry birds, he showed no willingness to acquiesce. Becoming more angry than before he beat up and down with repeated petulant snarling cries, very different from the occasional exclamations of the others, whom he attacked and harried at every opportunity, though he contrived all the same to pick up the lion's share of the food. The rest seemed afraid of him and put up no serious resistance, although he often cornered them and dealt as much punishment as he could manage. I noticed after one encounter he had to alight on the water and spend some time in disengaging from his bill the downy feathers which the vanquished had left adhering to it. All the birds, except the most practised and confident divers, are shy of voluntarily submerging when there is any risk of being attacked on coming up again, and the attitude of this pugnacious individual made the salvaging of the bacon-rinds out of the question. Not only that, but his fierce and persistent onslaughts caused the rest to retire as soon as I ceased throwing in bread: when that cause of disturbance was removed he

---

29. EMN: I have used the masculine for convenience, but the sex of course was impossible to determine.

easily routed the loiterers and with angry cries flew at and frightened away any other that alighted in his neighbourhood. In a little while he had swept them all out of his kingdom—birds as heavy and as well armed as himself—as thoroughly as a good sheep-dog clearing a field. Three times I threw in bread and collected a flock; each time he dispersed them as easily, immediately the supply was discontinued, and won back the full extent of his old boundaries. These, of course, were not indicated by any visible landmarks—on three sides they consisted of smooth water.

This was on 16th November; on the 17th we again spent some time observing this gull and playing upon his territorial susceptibilities. We found him still in occupation and on summoning up a crowd by throwing in bacon-rinds he was worked up to an excellent pitch of indignation, hurling himself fiercely into the task of driving out the invaders, whom he attacked sometimes at the rate of half a dozen or more a minute, and with such businesslike ferocity that once or twice when he got a victim well pinned down on the water it looked as if there would be murder done. Again the apathy of the assaulted birds was remarkable. Far from resenting this treatment they took only the mildest defensive action when there was no opportunity of escaping without showing fight. To confirm our ideas of his boundaries we moved along, distributing food as we went, so that the main flock was persuaded to accompany us. Gradually the warrior dropped behind and, after satisfying himself about the main body, retired to the centre of his territory, like a spider in an invisible web, and proceeded to clear off the stragglers. It was noticeable that this bird never molested or paid any attention to the mallards, however many of them entered his domain, though if a gull alighted or swam up against them, however unobtrusively, it was ejected without delay. I now observed that at any rate two or three other gulls round the eastern margin also held territory, but they lacked courage, and when confronted with an invasion their reply rarely went beyond raucous protests. One or two made half-hearted rushes at trespassers. At this time the main body was occupying the centre and western half of the Pond while a good part of the eastern margin was reserved by short-tempered individuals who could be seen swimming about alone, on guard. On the 19th, food being now rather scarce and the temper of the gulls ugly, we again made life a burden for the territorial warrior: this time it took him almost ten minutes to effect a complete clearance after food had been thrown in. He might have done it much more quickly if our remaining had not encouraged some of them to hope for more.

On 20th November we resumed the study, which began to grow more complicated. It became necessary to distinguish our original territory-holder as 'A'. He still held the same territory with equal success. We watched for a long time the territorial politics farther south and gradually unravelled the situation. In the beginning there was a territory about a hundred feet long extending from near A's boundary to the next little bay, including the water up to about thirty feet from shore and all the land in the rear. This territory was held by another fairly energetic gull, 'B'. Outside this was another territory-holder, 'C', who had no coast at all, nothing but open water. One or two others spasmodically and ineffectively asserted vague rights which were not recognised and which they forgot all about the next time there was a rush for food across the Pond. The strictest territory-holders never, or hardly ever, quitted their domains. B's and C's territories overlapped, causing considerable unpleasantness, while A's and B's were separated by a diplomatic little strip of no man's land. On this, apparently by a joint understanding, no outsider was tolerated; the arrangement worked very smoothly and it may not have been pure accident. B's territory extended far inland—right back to the trees. Though C had no coast the overlap made one suspect an ambition to acquire some at B's expense. The characters of the three were very different. A was a truculent die-hard, quite indefatigable, who never by any chance quit his domain. C was also a stay-at-home, but gutless and unenterprising. B owned an extraordinary territory, but he had a trustful habit of often dashing away to scramble for food across the Pond. We had with us a fair quantity of food, bread, bacon-rinds and—I forget what this was for—vermicelli, which we used at favourable opportunities to develop the situation or to precipitate a crisis. In this way we dumped a whole mass of vermicelli off the cape in B's territory, the effect of which was to summon a hungry invasion of strangers. B, unlike A in such circumstances, was inclined to accept the situation, and he did not take nearly strong enough measures to harry them. Among these strangers was one immature bird, who almost immediately after coming on the scene began to grow quarrelsome and attack the rest.

These did not resist, and his mood was clearly developing into an ambition to annex the territory when B, the owner, came on the scene and made a general clearance, expelling him among the other trespassers. He retired with very little protest, but only as far as the bay, a few feet off, where he began to drive off other gulls and usurp that corner of the territory. B was being kept very busy up at the far end of his territory, onto which the energetic ferocity of A was continually forcing trespassers to retreat. He came down once or twice, and I believe caused the immature bird, 'D', to retire a little, but still not over the border. Once the offender took flight he was satisfied and, by condoning his settling again inside the territory, which A would never have overlooked, paved the way to its usurpation. The immature bird D waded at the greatest possible depth round the cape in B's territory, driving off trespassers as he went, while B was still active up at the far end. B, after clearing that end, seemed uneasy and loath to provoke trouble with the usurper, a first-year bird, apparently, with a dark terminal bar to the tail. But by throwing in food at the psychological moment we managed to precipitate the conflict. D had been steadily gaining in courage and self-confidence; he now made it clear that he did not mean to be bluffed. B, who by this time had lost much of his nerve, weakly conceded his claim to a part of the territory, and confined himself to resisting D's efforts to expand still farther at his expense. The frontier at this stage was undefined and fluctuating; both birds were secretly a little afraid of one another, making tentative sallies but taking care to retire again before reprisals. In this way I think they felt the adversary's pulse, by seeing how much he would stand. D was the more enterprising and was not handicapped by the need for guarding extensive frontiers in his rear; in the end he gained all he claimed. The disputed boundary between B and C was therefore settled, on the whole in C's favour.

I noticed only one or two cases of territory-owners attacking one another, and only where their territories actually overlapped. The bird on the defensive, instead of retiring like a trespasser, would ward off the attack by crouching on the water with its bill pointed upwards. Boundaries were settled by a peculiar ceremony. When neither bird felt inclined to take the offensive both would alight, not face to face but side by side on the water, their wings loosely half-opened, necks crouching and heads raised (sometimes the tails also), both uttering the characteristic territory note in chorus and obviously as annoyed with one another as shopkeepers disclaiming all connection with the firm next door. This territory note, the 'Ouarrh-h-h!' previously described, seemed to be always and only used by birds holding or laying claim to territory. These were always bellicose and the rest, except other territory holders, gave way to them.

We gave a good deal of time to making and confirming these observations, for they give a glimpse of some unexplored regions of ornithology. In the first place it must be borne in mind that no black-headed gulls spend the summer on the Round Pond: therefore it is a case of seizing territory in winter quarters. This, I believe, is a thing undreamt of in the philosophy of the territory theory, as it has been stated by Mr Eliot Howard[30] and others. It is of course known that some birds keep territory all the year round, and Mr J. P. Burkitt[31] has found that some hens take up territory in autumn close to the breeding territory of the cock, but for a bird to assert such strong territorial claims in a place where it never spends the summer and does not even stay the night is an altogether different matter. It is true that most of the territories were impermanent, and varied from day to day, but some showed an astonishing continuity and A continued to defend his claim stoutly until he was frozen out by the great frost at the end of the month. He would certainly have held it much longer but for this physical expulsion. After the thaw the system revived, and at times almost the whole margin of the Pond was claimed by one territory-holder or another. There is also some significance in the fact that the black-headed gull is by no means an individualist species at any time: it nests in colonies and feeds in flocks. This unsocial behaviour may be exceptional or it may be merely a faint survival from an earlier stage of social organisation, but in any case it is hardly explicable by existing theories and the point is worth investigation.[32]

A few other simple experiments occurred to us. The black-headed gull is a very widely distributed species. He flourishes on

---

30. ED: E. Howard 1920, *Territory in Bird Life*.

31. ED: J. P. Burkitt 1924–1926, A study of the robin by means of marked birds. *Brit. Birds* 17: 294–303, 18: 97–103, 250–257, 19: 120–124, 20: 91-101.

32. ED: See D. Lack & L. Lack 1934, Territory reviewed. *Brit. Birds* 27: 179–199, 266–267; E. M. Nicholson 1934, Territory reviewed. *Brit. Birds* 27: 234–235; and C. B. Moffat 1934, Territory reviewed. *Brit. Birds* 27: 235–236.

the sea coast, on rivers, lakes and marshes, on pastures and ploughlands and shingly wastes like Dungeness and even on the moors. Yet it is possible to give a very wide and roughly accurate definition of the places he finds eligible: he may inhabit almost any place that is open, but he does not like to feel shut in. Therefore he avoids streets and woodlands and coasts overhung by lofty cliffs and all kinds of timbered and enclosed country. In London this prejudice has been very much relaxed. He will come to windows facing his regular haunts if food is obtainable: I have seen many feeding on the Embankment at a top-storey window at least sixty feet up, settling freely on the balcony, and I have noticed them coming for food to a balcony opposite Regent's Park lake. They will also, especially in the early morning, range over Kensington Gardens even among the trees, wherever it is a little open. I have seen them settled on the grass right over near Bayswater Road, and on the gravel drive of the part of the Gardens east of the Long Water, and also beating up and down near the Broad Walk, fairly frequently. But out of the open they are rarely trustful.

On 21st December, having amassed a vast supply of bacon-rinds, we gathered a swarm of gulls at the north-east side of the Round Pond and retired feeding them, to see how far they would follow. As soon as we had crossed the main diagonal path and left open ground behind us the majority deserted, although they were ravenously fond of rinds and we made it very clear that there were plenty left. Those which remained became suddenly very suspicious and refused to catch any more of it on the wing, nor would they descend and retrieve it when it fell on the grass, even after we had left it a long way behind. About a dozen hung on half way down the glade to Speke's Monument.[33] They followed us but that was all; their greedy tumult of a moment before was replaced by a silent suspicion and they flew considerably higher. Rather than come down for the rinds between trees and in such suspicious circumstances they went away hungry. This conspicuous example of their distrust of shut-in places—claustrophobia, to use the accepted jargon—tended to confirm my idea that a number of birds, particularly plovers and gulls, are so habituated to wide open

---

33. ED: A monument in Kensington Gardens, between the Fountains and the Round Pond, to the nineteenth century British explorer of Africa, Captain John Hanning Speke (1827–1864), who was the first European to see Lake Victoria; Speke's Weaver *Ploceus spekei* was named in his honour.

surroundings that a real fear haunts them in opposite circumstances like a child's fear of the dark. The skylark, wheatear, meadow-pipit, lapwing and various other species seem to suffer from it as well. It is difficult to explain otherwise how it happens that such specialised woodland and arboreal species as the bullfinch, hawfinch, green woodpecker, tits, wryneck, redpoll and others are comparatively often seen in open places and on the ground while the reverse—to see the typical birds of open country in woods—is so extraordinarily rare, even on migration.

This incident was typical of the character of the black-headed gull, which in London is a mixture of confidence and suspicion. In familiar surroundings, where neither prejudice nor experience has ever led him to doubt his security, he is bold to the point of insolence. On the Embankment by midwinter they will demand food almost with threats from anyone who stops to gaze over the parapet; at the Dell end of the Serpentine I have had them settling on my head when my brother was feeding them, and easily induced them to alight upon his own by putting a bit of food on his hat. Yet in the early morning mist, and also towards dusk, they are nervous and suspicious, swerving away in alarm if they suddenly come upon a man in the half-light, when their own forms appear strange and exaggerated—one by the Serpentine on an October morning looked for many seconds actually larger than a greater blackback.

Towards man, whose disposition, friendly in some places and murderous in others, must be a perpetual mystery, he behaves with a proper reticence, forgetting his precautions only where he knows himself welcome and resuming them at night even where he is most confident by day. With other birds he is much more sure of himself, and the relationship between the gulls and the remainder of the bird population is fascinating to watch. He commands the awe of the sparrow community, who with unfailing politeness refrain from touching any food on which a gull is seen to have his eye. Sometimes by collecting sparrows and gulls beside the Round Pond and by feeding the sparrows with large and crusty pieces which are difficult to divide or swallow it is possible to start a satisfactory chase, but the sparrows are poor stayers and if there is the slightest danger they let it drop in a panic. They had, however, a useful habit of bolting for a crevice in a pile of chairs, where no gull could follow. On the ground they were rarely threatened.

But it is with the ducks and other gulls that competition is most severe. On the Round Pond, for example, most of the food consumed is brought down and distributed by the Londoners. As they toss it up part drops on the ground and falls to the sparrows or shore-going mallards, part falls in the water and is eaten by ducks of all four species,[34] and part is caught on the wing by black-headed gulls. In addition to this the black-headed gulls generally obtain a good share of what drops in the water, either directly by picking it off the surface or indirectly by forcing a duck to give it up, and when the struggle for existence is particularly severe they also stand about and obtain a good deal of the part which falls on the ground.

The hierarchy is completed by the **common gulls**. These never, in my experience, accept anything directly from man: they hang about the rear of the swarm of black-headed gulls like scrum-halves and whenever some fortunate with an especially good catch attempts to retire and eat it in peace they are on his track in an instant. 'Waiting on' like trained falcons, they have almost a falcon's length of wing; at any rate with the long strides of their supple wings they overhaul the slower black-heads as easily as a pursuing hawk. With whining cries of excitement they strive for the position above and just behind the victim, where a falcon would dash in and strike. But here, fortunately for the smaller birds, all likeness to a falcon ends. They have not his strong foot to strike with, possessing instead a feeble webbed one, a blow from which would compare with a falcon's stoop like a smack from a clown's bladder with a straight thrust from the sword. The black-headed birds are too well aware of this ever to be bluffed into surrender by it, but, if they are not dangerous, common gulls are at any rate persistent, and this superior speed makes it very hard to shake them off. That is the meaning of all those interminable wailing whining chases which one sees in London almost wherever the two birds occur: the black-heads are reluctant to surrender to a bird which they know cannot really hurt them and, when they give in, which very often happens, it is from exasperation rather than fear. Even the pirates have larger pirates to torment them.

---

34. EMN: i.e. mallard, pochard, tufted duck and gadwall.

Apart from common gulls the **black-headed gull** in London recognises no peers. To the ducks, larger birds but without his masterful disposition, he is brutal and overbearing. The parasitism of black-headed gulls on tufted ducks has already been described by Mr H. J. Massingham and Mr Rudge Harding informs me that he noticed it on St James's Park lake in the first decade of this century. The habit all the same appears to be a new one, for I have not been able to discover any record of it outside the metropolitan area,[35] and here the gull at any rate has established himself only recently. The procedure shows little variety. Often a single duck is accompanied by two or three gulls, to whom he serves as breadwinner. Their method is to sail at leisure till they become aware that the duck beneath is ascending, when they at once take wing and beat over the exact spot, so that when he rises to the surface the silent menaces of unfriendly wings winnowing a few inches above him cause him to dive again in a panic, leaving his catch to be retrieved by the pursuer. But often the gull remains on the wing all the time, and perhaps takes an interest in more than one duck. Sometimes when he alights he holds his wings still erect and fully spread in order to be away at the first indication of a victim. It is the skua's method applied to another element. The surprising thing is that it is all done so quietly. Often neither the assailant nor the victim utters a sound. Because the tufted duck is by nature peaceable the trick is successful beyond its merits, and even if the victim dives he is liable to be assaulted time after time when he attempts to come to the surface. I have actually seen the black-heads plunge vindictively at one when it must have been swimming about a foot under water, diving with all their might at its clearly visible form. But if the duck refuses either to dive in a panic or to surrender his catch the gull is checkmated, and I observed that when that happened he generally desisted at once.

---

35. ED: There are several subsequent references of interest. R. Meinertzhagen (1959, *Pirates and Predators*: 56–67) quoted a note—entitled Black-headed Gull alighting on the back of Tufted Duck—by B. L. Sage in *Brit. Birds* (48: 177), published at a time when EMN was the journal's Senior Editor, and commented that he had 'seen similar attempts by the same gull to take food from tufted duck in St James's Park, but the duck always managed to dive again and avoid acts of piracy.' More recently, E. Gillham (1986, *Tufted Ducks in a Royal Park*: 34, 170–179) described competition between these two species in the central London parks.

On only a single occasion have I seen this piratical behaviour actively resented. That was by a young drake of the year on the Round Pond, on 22nd December 1925. The black-heads were exceedingly troublesome that day, but this young drake, though he had so far assumed no more of the adult dress than short plumes on the head, repulsed them by the same means as they use against one another. On being threatened, instead of attempting to escape by swimming or diving, he would face boldly round, raise himself into a half-erect position and, with open bill craned upwards towards the attacker, utter harsh notes of defiance and show such a bold front that even when three or four of them came on together they were always beaten off in the end. He used this noble art repeatedly, but the others still let themselves be imposed upon, and I have never seen an adult seriously attempt to defend itself.

On the Serpentine I have watched a tufted duck so harassed by black-heads that she was compelled to take wing, after repeatedly diving in vain. Even then she had some difficulty in outstripping and shaking off her pursuers. This savagery seems a part of the character of the species. On the Round Pond I have seen one single out a wild duck, who had got no food and was quite harmlessly employed, and badger her about most unmercifully, hovering threateningly above her head. She came on shore, but he and some others followed; she took wing but could not manage to shake them off. They continued to persecute her when she dropped onto the water and made her run the gauntlet, uttering all the time a curious petulant cruel cry, rather shrill and piercing. In the end they grew tired of their sport and gave it up.

At all times they are rough and boisterous, but as the dead season advances they become perceptibly fiercer in temper, constantly making harsh ejaculations and, when expecting food, quarrelling and attacking one another without provocation. The contrast of this behaviour with the meekness of the traditionally quarrelsome sparrow and the healthy excitement of mallards in similar circumstances is striking. Their bad temper is vented on other birds besides themselves: the instance already given was perhaps a depraved kind of sport, but on 19th November 1925, in the first week of the hungry season, we saw in a short time no less than six brutal and unjustifiable assaults on tufted ducks, which had not even any food to yield, floating harmlessly on the water. In most of these the gull actually alighted on the duck's back before it had time to dive and dealt it savage blows on the head.

A more humane and excusable practice, once witnessed by my brother when I was looking at something else, was this: he threw in a bacon-rind just in front of a wild duck, which craned her neck forward to take it, but a gull, swooping down from behind, alighted an instant on the duck's head, pushing it for a moment under water, snatched the rind from in front and was up again in a second. This was probably not premeditated, but if it ever became a regular stratagem it would prove most useful to the gulls. As it is they fully understand the value, in harrying a duck, of putting it in a panic from behind, and just above, the head.

On the Round Pond their air power dominates the situation. They could obtain almost every scrap of food which was intended for the ducks and sparrows, if only they wanted it badly enough. When they set their heart on a piece they generally get it, whether it has fallen to a land-bird or a surface duck like the mallard or an expert diver like the pochard or tufted duck.

It will be seen from this that it is not easy to distinguish between hunting for pleasure, sheer brutality for brutality's sake, and piracy for the sake of food. They almost overlap. But piracy in the purest form is practised not only by the common gull on the black-headed and the tufted duck, and by the black-headed on the mallard and tufted duck, but frequently on the pochard and sometimes—my brother[36] has noticed this—on the Serpentine coots. I have seen an unfortunate pochard forced to dive again and again, at least eight times, by two or three persecuting gulls. The habit seems to be spreading. I have seen it done freely as far out as Staines Reservoirs, and also on the little South End Pond on Hampstead Heath. The black-headed gull differs from other species which turn economic conditions to their own advantage in being less dependent on civilisation. He follows the plough, and anything else that holds out a hope of food, but he is not even indirectly a common parasite on man: all living things are his potential victims.

---

36. EMN: In Ayrshire my brother has often noticed them victimising lapwings in the fields, and Mr R. H. Brown (1926, Breeding habits of the lapwing. *Brit. Birds* 20: 162–168) comments on the same thing in Cumberland.

At the harbour of Combe Martin, on the north Devon coast, I used to be amused by the brutally simple order of precedence among the gulls when it was a question of enjoying the commanding perch on the post at the mouth of the bar. If a black-headed gull was in occupation and a stout herring gull came swinging up towards him he invariably deferred to the superior right of the heavy hooked bill with commendable smartness. Then if another herring gull came up with a determined front the occupier would grudgingly give way, not thinking the position defensible against an adversary on the wing. Only a very few times did the sitting bird show fight and succeed in holding the position. And when a greater blackback was seen to be heading for the coveted post no herring gull would dream of disappointing him of it. Everything seemed to show that the flying bird was acknowledged to possess an almost irresistible advantage. But the London gulls, always enterprising and original, have developed an art of defence which fully holds its own against the attack. On the Serpentine, when they are not engaged in scrambling for food, the gulls like to rest on the short posts supporting the chain round the island. The claimants often outnumber the posts by three or four to one, and competition for them is at these times keen. When a flying gull attempted to displace one sitting on a post the threatened bird would draw itself back into a fighting attitude, but instead of thrusting the bill forward would lift it directly heavenward, so that from whatever direction the enemy attempted to alight his dangling feet ran the risk of being trapped in the open beak. Generally it is a fear of the other gaining a foothold on its back that causes a gull to decamp, but this transferred the danger to the aggressor. The first time I saw it used was by a common gull, against a black-headed who had an ambition to settle on the post he was occupying. Instead of yielding he opened wide his yellow-green mandibles and turned them defiantly upwards, so that if the invader had carried out his intention he would have had to impale himself upon them. Actually he changed his mind. It seemed to be thoroughly well known among the black-headed birds, who used it often and effectively; those who failed had to balance uncomfortably on the connecting chain. When I watched there later, at the end of January, the weapons of attack and defence seemed much more evenly matched, so that the contest resolved itself into a trial of endurance. Second attempts were rare, and a displaced bird often flew straight to the next post and surprised its occupant, who, unless he had time to defend himself, had to decamp, and sometimes evicted the next beyond. Thus one eviction often resulted in two or three others. There was one especially complicated case. A moorhen struck a fighting attitude and drove a black-headed gull off a raft. He retired and captured a post, the displaced bird ejecting his neighbour, who thereupon attempted to capture a third, but he was driven off.

In the absence of vacant posts they will sometimes settle on a bush upon the island. Opposite *Peter Pan*[37] I have also seen one perched on top of a bushy tree about twenty feet high, apparently a thorn. It looked quite comfortable. The bare branch of a tall tree east of the Round Pond, always a favourite perch of the Kensington Gardens crows, also has a gull sitting on it occasionally, and one morning in mid November I saw two perched there together. Perching on trees has long been known as an occasional habit of this species; at such a tree-fringed colony as Scoulton Mere I have seen several perching together. The tall flagstaff of the Royal Humane Society's boathouse on the Serpentine was another point of vantage, the occupant 'laughing' whenever another gull approached. I have observed a herring gull make a very similar sound in the same circumstances.

For a web-footed bird such perches involve a skilful use of balance, but in that facility the black-headed gull is particularly expert. His all-round proficiency is equalled by few other species. Once, in an experiment on the Round Pond, we had the joy of seeing him impotent. While the Pond was frozen over, the gulls, which looked very grey and uniform against the reflected snow, crowded at the centre. They were very hungry, but when food was brought down and thrown on the ice they had to descend for it very circumspectly, and suffered many comic misfortunes, sometimes slithering two or three feet over the thin film of snow, sometimes falling over and losing their tempers in an absurd rough-and-tumble, sliding in all directions in their efforts to secure the rinds. Even in the act of alighting they were apt to reel about most drunkenly on their webbed soles. Here they were at a

---

37. ED: A delightful bronze statue of J. M. Barrie's character Peter Pan by Sir George Frampton situated on the western shore of the Long Water in Kensington Gardens. A photograph of the statue appears opposite page 21 of *London's Royal Parks: An Appreciation* by Richard Church (HMSO 1993).

disadvantage, but by no means baffled. But eleven days later, on 8th December, the thaw finally set in and their numbers at the Round Pond fell sharply. Men were sent by the Office of Works to smash the floating ice along the margin, which they did most thoroughly, and the next day we put to the test the gulls' amphibious capabilities, by attracting a hungry throng of them and placing bacon-rinds on the fragments of floating ice, to retrieve them from which needed great skill. Whenever a gull alighted to take the bait the ice beneath him promptly sank, leaving him either to fly off in alarm or to swim for it—we saw both done many times. If the ice tilted, allowing the rind to sink, the water was in too narrow a girdle and too much blocked with ice to give any hope of salving it. The behaviour of a bird when after gingerly landing he found the ice sinking under him was truly comic, and they could devise no satisfactory way of seizing the food, most of which went to the bottom, to remain there until the thaw was complete.

But the black-headed gull is not easily baffled. His ordinary dispositions are ingenious and scientific. Nothing could be more simple and efficient than the method observed by these birds when they are fed from the hand, or persuaded to catch food in the air, on a calm day at the Round Pond. They revolve in a figure-of-eight, having its crossover by the person feeding them, and consisting as a rule of one wide circle on the water side and the other narrower one over the land.

All the birds on both circuits—they change over continually from one to another—are invariably facing in the same direction when they come abreast. If there is a breath of wind they face into it. This is a very ingenious arrangement; it prevents any rough-and-tumble at the critical point, which would be inevitable if they approached as they liked and in opposite directions. As a means of dealing with numbers repeatedly passing a given point it is a vast improvement on the queue, though obviously it would work less smoothly on land than in the air. I have often singled out a recognisable bird from the procession and followed his movements: he would sweep rapidly round the great circle over the water, slow up towards the feeding point to loiter in the zone of windfalls for as long as possible, and once past the point would join sometimes the little circle, sometimes the great one, to revolve with as little delay as he could manage to the feeding point again. It occurred to me that it would be interesting to single out one distinctly recognisable bird and feed it alone, every time it came round, for the pleasure of seeing how much it would eat. It should be done early in the morning, to ensure an empty stomach. I tried something of the kind one day, but my bird was a fool and could hardly catch anything; his nerves also were bad and he shied off if it was not very gently thrown, or if another made a rush towards it.

Often in the early spring I used to bring down a good supply of rinds and march them all past, one after another, to examine at close quarters the progress of their developing hoods. It was unexpectedly slow. As early as 17th November 1925 a bird was noticed on the Round Pond which at a distance appeared to be completely hooded. On a closer view the head proved to be of a rather light sooty grey, almost white on the forehead and with a single darker marking. Even at close quarters it was nearer to summer than winter plumage. On the 19th there were two birds in this condition present, and the next day one of them showed a disposition to take up territory. On 2nd December I noticed one, perhaps one of these same birds at a more advanced stage, which seemed up to as close as twenty yards to possess a perfect hood. When it came nearer than that a few blemishes appeared. On 25th January, when I scrutinised them all, the great majority still showed no more trace of a hood than previously. But immature birds were rapidly assuming the grey, and a handful of adults showed well-developed hoods. One was chocolate-grey and perfect except for the tract in front of the eye (both throat and forehead) which was still very mealy. It was not until 15th February that I noticed a hood which could strictly be called perfect: I examined it closely by throwing food to that bird alone every time he came round out of all the clamouring horde, and so causing him to hover for inspection in front of me. Even by 3rd March not more than 30 out of 176 birds on the Round Pond had hoods. In other years I have noticed the majority changed earlier, but up to the last week in March, when most of the adults have departed, the assumption of the summer hood is not complete.

This is one of the privileges which only the bird-watcher in London can enjoy. There are occasions elsewhere, especially on the coast or marshes, when one would give one's right hand to be able to call up birds for close inspection in the same way.

Often I scrutinised their dangling scarlet feet, hoping to find a ringed bird among them, but I never discovered one. It seems to me a pity that the marking of the black-headed gull in this country should have been given up: a species which can produce two transatlantic records against such heavy odds is not without interest, and we can scarcely claim that our knowledge of its movements is complete. It is not even known, apparently, where these London birds breed. They may belong to East Anglian gulleries, but that all are from the same part is unlikely. It is known that native British birds wander abroad, especially to France and Spain. Also, there are certainly many natives of the Baltic countries along our east coast during the winter. Danish birds have lately been recorded from the Humber, the Wash, Colchester, Brightlingsea, near Rochester, near Folkestone, Rye and Glamorgan, among other places; others from the Baltic islands were reported in Lancashire during November 1923 and at King's Lynn the following winter. In the last two months of 1925 north German birds of the year reached Gorleston and Great Yarmouth, and an Estonian marked specimen was reported from Lowestoft. In February 1924 a bird from Utrecht was found in the Thames Estuary and one ringed in Sweden on 15th June 1925 was reported from Walthamstow, actually within the metropolis. The probability is that the clamorous swarms which we feed in London are neither purely English nor all strangers, but a cosmopolitan crowd of English, Scottish, Dutch, German, Swedish, Danish, Estonian and Lithuanian birds. It would be interesting to find out, and the catching and ringing of the London wintering birds should not be difficult in hungry weather, if the authorities would allow it to be done.[38]

Often there were present birds with large burnt-looking or chocolate-coloured patches on the lower breast. The appearance of these scars on close inspection and the exactness with which they corresponded to the keel-surface of the bird when it swam—like the water-line of a ship—left little doubt that these were the victims of floating oil.[39]

---

38. ED: For details of recent ringing recoveries, see page 168.
39. EMN: William Beebe describes in *The Arcturus Adventure* (1926) how a mysterious gull with jet-black breast and underparts was sighted flying high overhead on the Sargasso Sea. He shot it and found it a kittiwake 'in good condition but with the ventral plumage saturated with oil, into which it must unwittingly have swam'.

In a high wind, or when they were very hungry, the figure-of-eight was sometimes abandoned. Wind destroyed it, because with a sufficient current of air the birds at the feeding point could hover indefinitely, without ever needing to circle back. They were masterly in the use of air-currents, and nothing delighted them more than a windy day. In a stiff breeze I have actually seen one fly backwards, by checking his impetus till it was appreciably less than the contrary speed of the air. Their management of the currents and eddies about the bridges over the Thames is marvellous. But the most spectacular display of their skill occurs when they make their daring descent from a height. Then they will often plunge at breakneck speed, urging themselves faster by great strokes of the wings or holding them bent back like a bow and more or less depressed. The swerves and hurtlings become too quick for the eye to follow, but one of the most characteristic is a lightning roll, the bird careering along with its wings banked, for example, at five minutes to five by the angles of a clock, and instantaneously flashing over on to the other curve with its wings at about five past seven. The feat is often repeated several times in quick succession, and frequently in horizontal flight just above the water. This lightning reverse does not involve turning on the back, but that undoubtedly occurs during the rapid descents.

One windy day in February I saw a black-headed gull hotly pursued, rush down with immense impetus and loop the loop perfectly, overhead, finishing within ten yards of me. The acceleration, plunging with the wind, was simply terrific, and the bird described a fair ellipse, losing speed rapidly as it travelled upside down over the top. I have seen various large birds, including the buzzard and heron, perform somersaults when descending from a height, but looping the loop in cold blood, as an aeroplane does it, is a very much rarer and more difficult feat. Sometimes even the hurtling switchback flight is done at a few feet above the water; occasionally it resembles the staggering of the lapwing, but more often the swerving flight of a snipe. Always there is a hint of its relationship to the waders in this idiosyncrasy: the green sandpiper behaves like that and at times other sandpipers will do the same, but it is not the flight of a gull. Usually this extravagant behaviour seems due to sheer elation of spirits; rooks on a windy day will act in the same manner.

The London Gulls

Undoubtedly the weather affects them deeply. On some occasions they are shy and impatient, on others curiously silent. Rain scatters them broadcast, and if it does not reduce the population at the Round Pond at least it renders it more floating, parties coming and going continually and finding very little to eat. They are hungry and rather despondent because so few people come and feed them in the rain; it seems also to have a surprising effect upon their plumage, and on a moist warm rainy December morning I have found them so draggled and forlorn that their wings made a curious loud erratic noise when they first rose. This makes them more truculent than usual. On the bitter morning of 4th December 1925, with the Round Pond under three inches of ice and a thick fog which had settled the night before limiting extreme visibility to about twenty-five yards, the black-headed gulls failed to put in an appearance, or perhaps saw the state of things and retired elsewhere. There was no trace of them in the middle of the morning; normally at this time of year they arrived about half-past seven.

There is scarcely any part of Greater London where they cannot frequently be observed, particularly in unsettled weather. Even above Tottenham Court Road, Oxford Street, Bond Street and Park Lane I have often seen them in flight. Their actual haunts are more limited, but there is practically no sheet of water, except the very smallest, which they do not frequent, and I have seen them feeding country style on fresh ploughland as close as Barnes. Yet they are abundant only on waters where the public have access and supply them with food: on reservoirs the gull population is by day comparatively sparse, and though they penetrate often to Paddington wharves there are long stretches of canal for which they show little liking.

The **common gull** is much less plentiful, but has nearly as wide a range. It is, however, more numerous than any other species, except the black-headed, and few of the chief haunts are without a handful of commons. I have seen one passing over New Bond Street. In general behaviour it is similar to the black-headed, but far less versatile and less trustful of man. The characteristic cry, a prolonged almost whinnying snarl, querulous and discordant, often proclaims its presence before it can be picked out. In severe weather the numbers on the park lakes are enormously increased.

By all accounts it appears to be much commoner in London now than it was not many years ago.

I have seen common gulls attempt to plunge in the same way as the black-head, and in a high wind I observed one walking on the water with wings full spread and dangling feet, as the storm petrels are said to do. The black-headed gull will do the same. They are inordinately fond of chasing one another, and enjoy overhauling another and wresting his food from him much more than obtaining it with less exertion for themselves. If there is no food, to continue the game they will play it with a substitute. I have watched one on the Round Pond pounce down into the water and pick up a large twig, bearing it off at full pelt with others in pursuit. The bearer, being overhauled, let it drop, but it was snatched up and captured several times more before the game was given up. Undoubtedly it was a game, for the twig was very obviously wooden and all must have known perfectly well that it was not anything to eat. I have seen black-headed gulls play the same game with dead leaves.

Chapter III

The Flylines over London

Towards sunset, every day between the middle of October and the early part of March, a very general and conspicuous movement takes place amongst the bird population of London. Within the space of forty or fifty minutes many thousands of birds perform journeys from their feeding-place by day to the roost where they spend the night in company. These journeys, though often brief, may extend as far as ten or twenty miles. At first sight they seem bewildering and impossible to reduce to coherence: everywhere from Hampstead to south London and from Walthamstow to Richmond one sees birds in flocks or parties hurrying across the sky, and it seems a superhuman task to map their little migrations and discover where they are going. But, complex as they undoubtedly are, these movements have a definite and simple object, and once a satisfactory theory has been found there is no serious difficulty in tracing out the whole of them, beyond the amount of laborious field-observation required to survey so great an area of house-covered wilderness during the brief hour or two each afternoon when the movements are taking place. I do not pretend to have worked the subject out completely. That in fact would hardly be possible for, though the roosts and flylines are regular, they are not absolutely fixed and unchanging, and by the time the most elaborate investigation was finished some part of its findings would already be out of date. What I have done is to observe and to make theories and then to observe again till they were either proved or found wanting, not tracing out every similar case in cold blood, once the general principle had been discovered, but turning to the next problem. At the end, therefore, I had my detailed investigations of each marked on the general map and in addition a great deal of other movements noted incidentally but not worked out in detail unless they seemed to show some discrepancy with what had already been discovered.

These movements, certainly, are by no means restricted to the metropolis. They occur in all parts, and the metropolitan system is at its outer extremities inextricably entangled with purely rural

The Flylines over London

movements. In Buckinghamshire the year before[40] I had been interested in the redwings which congregated at dusk in the middle of a beech-wood in the Chilterns, journeying there in company by a regular line of flight from the levels of the Thames where they used to feed all day long. The rooks at this place had a flyline several miles in length. Ducks and geese perform similar movements on which wildfowlers often rely, and in fact all birds whose best feeding-grounds do not practically coincide with the most favourable roosts exhibit something of the same kind. But in London various peculiar conditions, especially the quantity of houses, which are barren land, and the abundance of cats, lessen the possibility of a suitable roost being available near the best feeding-grounds, and exaggerate the scale of the migration. The movements are also more easily noticed where the standing population is so uneven, and the very nakedness of the streets gives any sign of wild life a fresh significance. Here we become keenly aware of sights and sounds which in the country are as common and as lightly accepted as night and day. The field of sight is oppressively limited: in London one lives and moves in deep trenches, eternally shut in by buildings, and the eyes, accustomed to the liberty of a more ample horizon, turn for relief to the open spaces of the sky.

To describe first the movements of the **black-headed gulls**.[41] The gulls we see all day long in central London do not roost there. Thus confidence, as among other species trustful of man in London but more shy in the country, is born and dies with the daylight. The same gulls which snatch food from your hand at noon will start back in alarm if it is openly tossed to them early in the morning or at dusk. They could not feel themselves secure if they slept within the orbit of the metropolitan din: at night, when their sharp wits are put aside, they must be outside the pandemonium again in their own sane world.

I was aware, in a vague way, that they deserted the Serpentine at night. When the subject began to interest me I watched and found that they flew up it about sunset to the bridge and there

---

40. ED: i.e. in winter 1924/25.
41. ED: See also R. W. Hayman and H. J. Burkill 1926, Roosting Place of London Gulls. *The Field* 147: 172.

wheeled to the left towards Kensington Palace. One afternoon in the first week of December (1925) a watch was kept at the Round Pond. Long before sunset, even before the light began to fade perceptibly, the first gulls were seen flying westward over Kensington. These were clearly the birds which had been observed on the Serpentine, rising one by one and flying in this direction. They came over the Round Pond at a height of three hundred to five hundred feet, well in their stride and using that easy swinging beat which is the gull's marching pace, sometimes singly or in twos and threes, sometimes in straggling skeins or V formation. Underneath they saw the Round Pond gulls still noisily feeding, and not many could resist the temptation to descend and join the fray. A few, mostly those too far out on the left wing to catch sight of the lure, swung unhesitantly on; others glided down but changed their minds and passed over. Many came pitching the whole way down to earth and mingled eagerly with the cloud of wheeling birds catching tossed-up bread. There was, however, a contrary movement. The gulls of the Round Pond must also be going home, the benefactions of bread notwithstanding, and hardly a squadron passed over without determining some of those beneath to follow while the light lasted. Occasionally, too, when a gap in the succession of bread-givers brought a lull in the frenzy, a bird would rise and wheel round with an inviting 'Wouw?'—a soft and interrogatory version of the more drawn-out territory-cry—which had a clear equivalent in our own language such as few bird-notes have. It signified 'Who's for home?' or simply 'Come'. This sound, which I afterwards came to know well, was fascinating in its expressiveness: to hear it suddenly amongst the babel from the scrambling birds was like hearing a remark spoken in plain English in the middle of a Chinese riot. To the unaccustomed ear it might have sounded unlike the rest of the gull-clamour only in its isolation and its greater emphasis, but having found the clue I was always afterwards alert for it, and I have never yet heard it uttered without proving the signal for departure.

In response to this call a party of from half a dozen to twenty gulls would take wing deliberately, the proposer wheeling into place amongst them, and with steady business-like beat they would rise over Kensington Palace and disappear westward. Many times afterwards I watched the procedure at the Round Pond. It varied somewhat with the weather, for though on chill midwinter days when the sun set early they would get up in little groups and fly doggedly away without ceremony, they gained great benefit from the lengthening afternoon, not lingering till dusk when that became later but departing at their leisure by broad daylight. When sunset was at 3.53 p.m. it might be nearly quarter past four before the last stragglers had torn themselves away from the food with which late-coming benefactors continued to tempt them. Two months later, there were days when I could come down to the Pond in broad daylight at four o'clock and find not a gull remaining, so that one had to look many times, carefully, to discover what it was that made the place seem so naked and derelict. Since October they had been present constantly from dawn to dusk. On only one occasion—the morning of the bitter 4th December, when the Pond bore nearly three inches of ice and a dense fog on top of it—had I missed them there by daylight, and their white forms had become an accustomed part of the scene, without which all appeared unexpectedly lifeless.

On these lighter afternoons their departure became more leisurely and ceremonious. At a much greater height than formerly—so high at times that it was hard to pick them out—other flocks and parties of homeward-bound birds arrived above the Pond. Seeing what was going on beneath, and having now plenty of time before them, they began to wheel round and round in circles, till by degrees one bird after another would succumb openly to the temptation and come tumbling down at full pelt to join the riot. At first the numbers seemed to increase, but very soon the homeward movement began here also, and the departures of independent contingents or of birds joining passing flocks outnumbered those coming in. This ceremony was, especially in fine weather, far more graceful and less peremptory than before on the short days. The note of invitation would be taken up by half a dozen birds, flying round in circles over the water, appealing to those underneath, till there would be formed a whole whirlpool of circling gulls, very gradually ascending as they went round. This whirlpool (if a whirlpool could be so calm and leisurely) did not, after it had reached a moderate height, continue over the same place: still revolving it shifted westwards, so that the centre was first over the water, then over the grass, then over the trees of the Broad Walk and, in this way, still eddying and rising, it was lost to sight over Kensington.

But before this time I had followed them out and discovered where their roost lay. Leaving Kensington one afternoon as soon as the first parties came over I got on top of the first bus to Barnes, towards which the line of flight apparently pointed. On the way I saw more of them pass over and observed with satisfaction that, though at first they were flying north of the road, they crossed it near Olympia, steering in the right direction. At Castelnau I got off and walked along the right[42] bank of the Thames towards Barn Elms Reservoirs till I came in a few minutes to the lower tanks, of which the nearest was filled with swarming gulls making an indescribable clamour. This babel was altogether different from that heard at the feeding-places, though on the Rhine I had listened to it from a flock hunting over a shoal. It reminded me to some extent of the starlings' roosting chorus, mixed with the grating note of the mistle-thrush, and others like the quack of a duck, but though the individual contribution, an abrupt grating note, may well have been louder, the volume of sound was smaller, and it carried badly. I now began to see gulls coming in across the River from the direction of inner London, mostly in squadrons of twenty or forty, but sometimes less and sometimes a hundred or two at a time. Walking the whole length of the reservoir I still found gulls pouring in on a front of half a mile or more, mostly from the middle of London but some from the east and even the south-south-east—from Putney and Wandsworth—and others from the west. The total number that I saw come in was certainly to be reckoned in thousands. The spectacle of their retiring was magnificent: they appeared in flocks, flying at a height but plunging in a whirlpool of circling birds as soon as they saw the river underneath; then checking themselves, still at a fair height, and making a leisurely slanting descent, slipping through the air in a flung-out line, utterly silent, their wings bent like bows, till they came gently to rest on the water and relaxed themselves at once for sleep.

The ghostly silence of their return in the twilight, their fine spread of wings and their surprising numbers—an unbroken stream of up to three hundred a minute—make this homecoming of the gulls about the shortest day one of the impressive sights of London. Understanding the importance of leaving not a shadow of doubt I scaled the spiky railings with the help of an overhanging

---

42.   ED: i.e. south.

branch, crossed the embankment and lay watching with field-glasses on the shaggy grass by the partly frozen reservoir till darkness came and all but the latest comers had settled down for the night, for which felony may the Metropolitan Water Board forgive me. As things turned out this absolute proof that the London gulls had slept that night in thousands upon Barn Elms was a vital point in the inquiry.[43]

From the Round Pond to Barn Elms is only two and a half miles as the gull flies. What I had seen with the intention of satisfying my curiosity had served only to increase it. Where did all the other squadrons come from? Since the Serpentine birds followed every evening a single narrow well-defined route, might not these also? Was it after all practicable to map their lines of flight, to find out definitely that the birds of a given species always roosted at some other given place and travelled there by regular flylines? From that day on I began to mark carefully on maps any movements that were observed towards roosting time, and to contrive to be out of doors at that time every day when the weather was favourable, come whatever, however much work I might have on hand. Within three months I had all the details I wanted except for the outer suburbs, but to record them all in the order in which they were actually discovered would be confusing, and here each species is treated by itself.

First it is well to finish with the gulls. I do not deal specifically with the common gull as distinct from the black-headed, since so far as I could ascertain the behaviour of the two species in connection with roosting was everywhere precisely the same. They have their roosts and their flylines all in common; they travel at the same times and often in company; and even in small details I could find no discrepancy except that the common gull appeared not to possess, or not to make much use of, a special roosting cry. This surprised me a little, and I took some trouble to investigate since the conduct of this species in London, described in Chapter 2, is not merely different but ought inevitably to breed hostility.

---

43. ED: In his 1975 book *Birds of Town and Suburb*, Eric Simms quotes C. J. Cornish's description (from *The Naturalist on the Thames*, 1902) of the scene at Barn Elms in the early 1900s. It was, he said, 'as sub-arctic and lacustrine as on any Finland pool, for the frost-fog hung over river and reservoirs' while in the centre of one of the stretches of water an acre of apparently heaped-up snow 'changed into a solid mass of gulls, all preparing to go to sleep'.

The Flylines over London

I had seen at Barn Elms the mass of the flocks come in from Chelsea way, with the Serpentine contingent on their right wing. There seemed only one place where such numbers could be raised and that was along the River, particularly off the Embankment. At the next opportunity I waited near Blackfriars Bridge and surprisingly early the hoped-for movement set in upstream, flocks coming in from London Bridge at a good height and on to Cleopatra's Needle, where the Thames swings sharply to the south, at right angles to its general direction. Here the main stream of birds followed their senses rather than the River and headed boldly overland. I felt that I could count on picking them up next time about St James's Park. This apparently safe inference gave unexpected trouble later. Three times I cast around for the birds at the right time around Trafalgar Square, but it appeared in the end that this route is an alternative, and not the one in everyday use. When they do not travel that way gulls continue up to Westminster and here the alternative route wheels to the right over the Houses of Parliament and the Abbey to the neighbourhood of Victoria. It is in this area that the flylines are most struggling and uncertain: they occupy a corridor at least a quarter of a mile in width and the course of the passing flocks is more often serpentine than straight. The confluence of the St James's Park lake birds is an added complication. These, though on fine days they gyrate upwards in a whirlpool like those described at the Round Pond, make sometimes in dull weather a most impressive departure, swinging off in a long column over Buckingham Palace and joining the main body between Victoria and Sloane Square.

They fly as a rule quite low over the house-tops and continue without incident over Chelsea till they strike the River at Fulham Palace or beyond, having cut off about two miles. Others, as I had suspected from the first, follow the River all the way and are joined above Battersea by the contingent from Battersea Park lake. This, and the details about Putney and Walham Green, were ascertained for me by my brother, whose skilful help made it possible to establish much that a single observer could not have found. The remotest sources of the flylines are undoubtedly beneath Tower Bridge, for watching there before sunset I have seen considerable flocks come swinging up from the Pool of London. Yet, on a later occasion, and so late in the afternoon that they must certainly have been making for the roost, we watched quite a number of odd birds and parties come down from the direction of London Bridge, some striking a little inland over London Docks and then continuing *down* the River.

The route from the Serpentine to Barn Elms, however, has as a tributary a long and remarkable flyline which has its origin in the Hampstead Heath Ponds but receives its chief accession from Regent's Park lake, passing from there right across the boroughs of St Marylebone and Paddington to the Water Tower on Campden Hill, at which landmark it swerves rather sharply to the south and joins the other beyond Holland Park. The Hampstead Heath contingent drawn from the ponds at South End is not a strong one and until it touches Regent's Park the flyline is straggling and insignificant. Even this is used by the common besides the black-headed species: it seems that the first will follow the second, on which it is in London largely parasitic, wherever it regularly penetrates. There is only one special point to be noted in connection with this flyline. From the middle of the morning on until the time when the flylines are in operation there is to be noticed over central London, especially well above Oxford Street, a constant leakage of gulls southwards towards the River. It is scarcely more than a trickle, rarely more than three or four birds being noticed at any time, but it takes place over a wide front and goes on for a long time, so that in the entire absence of a compensating northward movement the loss to the northern gull population must be large, and this probably explains why the Hampstead Heath and Regent's Park flyline is not so strong as it might be. A considerable number of gulls evidently perform a triangular daily journey, from the roost out to the northern suburbs in the morning, then across to the Embankment and back at night by the River.

By the middle of January this inner London system was fairly thoroughly mapped. It was clear that variations took place from day to day, so that a flyline recorded as following the course of a certain road might sometimes go a little to one side of it, sometimes to the other, but on the whole the marvellous thing seemed that they should be so little erratic and that, at the hour when these conspicuous squadrons were traversing London from end to end, it should be possible, away from the flylines, to go mile after mile without sighting a solitary party which had left the beaten track. With the starling and woodpigeon flylines it was not the same. There were certainly definite lines, and magnificent ones, but they came to some extent from all sides and there was no place where one could not expect to notice a few wanderers. The reason for the greater concentration of the gulls was obviously their attachment to water: always at roosting time the flocks would be journeying from one sort of water to another and their sense of direction and economy of effort kept them approximately to the straight line which they knew without Euclid to be the shortest distance between the two points.

But there were two disturbing uncertainties to be tackled, both of them made difficult problems by the fact that they concerned the more distant outposts, where transport was less convenient and mobility restricted, and where also the time occupied by this enquiry was at least doubled by the long journey to the scene of operations. It was one thing to get in eighty or ninety minutes of field work at the proper time each day and quite another to contrive to spend whole afternoons upon it. The first of these problems was the hypothetical Walthamstow roost, suggested by my brother, which had been allowed for on the map illustrating the working theory almost from the first, but never substantiated by observation. On 19th January (1926) we went out to Loughton Marsh, noting gulls in strength on the sewage farm and flooded fields, and made our way to Coppermill Lane, a long and curious thoroughfare running east–west between the Racecourse Reservoir (the most northerly of the Walthamstow group) and the slightly smaller one next to it (No. 5), commanding a good view. This we patrolled almost till dusk, believing that here we should interrupt the line of flight of all birds coming in from East London to the Lea Valley reservoirs. The result was disappointing. The influx from the south was meagre and from other directions we could see none at all. Later I discovered a line from Finsbury Park and Clissold Park which appeared to head to Walthamstow, but further efforts to unravel movements in the east were frustrated, as will be explained.

The second problem was of a more shadowy nature. I had proved in December that the London gulls roosted on Barn Elms and had imagined that subject to be finished with. Obviously there were suburban movements—in November, for example, I had watched a considerable influx of gulls at the Staines roost, presumably from the north, appearing to come from the River Colne. But these I did not propose to investigate. It was necessary

The Flylines over London

to draw the line somewhere, and I was reluctant to involve myself in cold blood in the investigation of any flyline which did not seem to begin or end in the heart of London. But the Barn Elms roost, upon which so much had depended, altered the situation. In January (1926) an anonymous correspondent of *The Field* stated that every evening gulls were to be seen flying south-west over Richmond and enquired where they were going. If only they had been flying east or north or south or west I should not have needed to concern myself with them, but that they should fly south-west obviously endangered the Barn Elms theory, for that is the way from London. Certainly I had seen them compose themselves for sleep at Barn Elms with my own eyes and it seemed unlikely that the reaches of the Thames between Hammersmith Bridge and Kew would provide a respectable number to pass over Richmond, making, as previous experience placed almost beyond doubt, for the reservoirs between Hampton Court and Sunbury. But the point deserved to be cleared up. Their departure from the Round Pond so immensely long before dusk during that January and early February went further to rouse my suspicions and to demand an investigation at the first opportunity. That, however, was slow in coming, and meanwhile I had been told by Mr H. J. Massingham that streams of them passed over Barnes Common coming from Barn Elms and going apparently in the direction of Richmond. This made it imperative to open the whole question over again as if the Barn Elms roost had never been proved.

An investigation of the peninsula on which Barnes stands revealed two main lines of flight, one coming from Hammersmith Broadway, greatly reinforced above Hammersmith Bridge, and from the group of little reservoirs opposite Chiswick, which crossed the River some way below the Railway Bridge at Barnes; the other coming right across the neck of the peninsula over the Red Lion. Both were very strong and, on this occasion at least, very high. I now hurried back to the towpath in front of Barn Elms, in order to observe what change, if any, had taken place in the procedure since December. The flocks were coming over from Hammersmith and Fulham as they had done before, but the numbers seemed to be less and there was this striking difference: instead of checking themselves and gliding or plunging down as soon as they saw the River, they passed on at great height without even pausing in their stride. Some flocks, particularly the latecomers, were flying lower and seemed strongly inclined to settle, and many disappeared below the embankment. A sense of moral rectitude, and the presence of an unpromising policeman on the towpath, dissuaded me from committing a second felony to get a view of the water. The position, however, was pretty clear: in December the shortness of the days had compelled the majority of the London birds to sleep on Barn Elms, which is the nearest feasible base to the centre, but, as sunrise became earlier and sunset later, more and more passed on, as the suburban birds hereabouts had always done, to the larger reservoirs up the Thames.

It remained to follow the flyline out and to discover where it ended. There had seemed no doubt that from Barnes, where I had left it, it went straight to Richmond and might be picked up near the bridge, but I had been deceived before and was not inclined to take any chances. I determined to go down the next fine afternoon to the western reservoirs at Barnes and leave with the earliest gulls. It was obvious that such a lengthy flyline as this one promised to be could not possibly be followed on foot, while other transport in the proper direction was uncertain and at the mercy of traffic hold-ups, which might spoil the observation of the line at the critical moment. There was nothing for it but the bicycle: a means of transport which, in London at any rate, is to be kept for the last resort. Saturday 13th February 1926 was the first fine day for a miserably long time. In the last fortnight the sun had failed to appear at all on eleven days out of fourteen and the total duration of sunshine in central London during the whole of the period was stated to have amounted to three-quarters of an hour. The opportunity was too good to be missed. Before half-past three I saw a body of gulls rise up in circles above Castlenau and set off south-westwards. I followed through Barnes and Mortlake, beyond which more gulls were flying abreast on the right.[44] Along the Lower Richmond Road to the depressing outskirts of Richmond, where a magnificent spectacle was presented—a great flock of gulls, on the left of the road, spread across the sky in a flung-out arrowhead formation, flying with the sun full on them at a vast height. They were

---

44. EMN: A more interesting event at this point was the appearance of four majestic swans, flying at a fair height in the same direction: they passed over a football ground with a match in progress. In some parts, for instance in the neighbourhood of Bournemouth, it is a common sight to see them on the wing, but in London I had never before enjoyed it.

heading for Richmond Bridge. There I stopped to look around. The flightline, it appeared, was rather broad and vaguely defined but the centre of it passed almost directly over the bridge and followed the same alignment. It was obviously set to continue through Twickenham, which I did too, noting flocks all the way, at first on both hands, later all on the left. After Eel Pie Island the main road left the River and the flying fell right away from it. Turning through Strawberry Hill I recovered it at Hampton Hill, where it came over in strength, recrossing the road at the bridge over the Cardinal's River. At this point it almost coincided with a strong flightline of rooks, bound for Kempton Park. Plunging into an apparently nameless suburb—they ought to give it a name: then there would be something distinctive about it—I found for the first time a fork in the route, one rather weak line going off almost due west, the other continuing north-west, as I did to Hampton Station. Here the flocks were plainly, almost ostentatiously, checking their steady beat and beginning to glide down towards earth. The end of the journey was evidently near. Having reached the Staines Road I unfolded the map and found that the apparent point of descent coincided with the Molesey group of reservoirs over the River and there was no bridge nearer than Hampton Court two miles away. This would have involved an addition of six miles to the journey: the problem was whether the roost could be proved without it from the Middlesex bank, for in these matters it is never safe to jump to conclusions, however probable they appear. I found a muddy lane leading to the lower road and, from two or three points, was able to observe with field-glasses flock after flock gliding and plunging beneath the embankment skyline of the large Walton Reservoirs, the south-western part of the group. This was conclusive: here was a main roost of the London gulls, if not the most considerable of all at this season.

The scene above this part of the route was marvellously impressive. Over the north-eastern horizon there would appear a bold slender V hairpin silhouetted black against a blue ground, as plain and formal as the chevrons marking the movements on my map. As it came swinging across the sky the flock grew larger and the pattern changed, as first the separate birds could be distinguished and then their flickering wings beating steadily like oars. Flock after flock came over in this way, all beginning to glide down as they approached with the sinking sun full on them, relaxing their discipline and breaking their ranks as they reached the end of their journey. One flock was like an ogee arch. Below them at this point passed an important flightline of rooks for Kempton Park and, where the black birds crossed the track of the white, I noticed occasional disputes over the right of way. Even intrinsically the sight was more impressive than I can describe. To me it was many times more marvellous to stand at the end of the road across the sky and think of the places to which it led: how some of the contingents were coming in by that vague trail over Regent's Park from the Hampstead Ponds, others from the Serpentine and others from the Pool of London; how they crossed out of London into Surrey at Castlenau and over into Middlesex south of Chiswick and back into Surrey at Mortlake, again into Middlesex at Richmond Bridge and once more over the Thames into Surrey as they descended into Walton; how over Regent's Park the starling lines to the British Museum and St Paul's crossed over at right angles and how at London Bridge the little black birds and the large white ones were flying side by side while at Castlenau and Fulham the starling stream was flowing in the opposite direction; how at Chelsea the town woodpigeons crossed over on their way from Hyde Park to the Battersea roost; and how near Kempton Park rooks passed them which had spent the day in Surrey fields outside the remotest tentacles of the suburbs. The knowledge of these ramifications, some of which I had discovered through deduction and others stumbled upon by luck, was a background that diminished in one sense the significance of the visible influx but magnified it in another.

Turning north towards Sunbury Common I recovered the northern branch of the flightline above the railway bridge near Kempton Park Station. It was not so heavily used as the other, but while I watched several flocks came over at a height, their direction not quite coinciding with the alignment of the railway, and with field-glasses I kept them in sight for I should think a mile or more, during which they did not change from their course which must have taken them to Littleton, but that roost was never confirmed. Other flocks were flying at a much greater angle to the line and appeared to be heading for Staines.

But the time was now approaching when the gulls would leave London and it was imperative that before their departure something should be done to tackle the still unsolved problem of the eastern roosts. Accordingly the next opportunity was taken for the exploration of the Lea Valley group of reservoirs which, though carried out as carefully as any of the previous observations, proved a dead failure, resulting only in getting evidence that a self-contained roost existed at the time on the King George Reservoir at Chingford, the birds which spent the day there remaining to sleep. On 1st March I made a further attempt to solve the problem by going down before sunset to the Iron Bridge over the Lea by East India Docks, the boundary between London and Essex, where it seemed highly probable the day movement towards Walthamstow would easily be observed. Finding only one or two gulls in sight I made desperate efforts to penetrate to the Thames—in this part of London it is extremely difficult to get access to the River—but, after much ineffective wandering, found a curious passage between high walls which led down to some waterman's steps and commanded a narrow and not particularly interesting vista of Bugsby Marshes on the opposite bank. Here I saw considerable flocks passing downstream at a fair height and in a direction which now definitely indicated the presence of a further unknown roost. Here obviously lay the flightline the first beginnings of which had been observed in the parties passing downstream at dusk by Tower Bridge. But the next afternoon, standing on Waterloo Bridge, I noticed hundreds of them rise loosely in circles—it was a fine day—and all without exception made off downstream. The old flightline was reversed, unless as one observer has suggested these downstream flows are entirely departures for the breeding grounds and do not intend to return. On 6th March the reversal of the main lines of flight had made itself felt as far as the Round Pond, but not decisively, so that at sunset I saw the extraordinary spectacle of one steady stream of birds coming up from the Serpentine and over the Round Pond westwards and another, as large or larger, going down along exactly the same line from the Round Pond over the Serpentine eastwards to join those passing down the River.

After the gulls, **starlings** show the most spectacular movements. These in fact are the greatest of all in point of numbers. Practically every church steeple and every great building from Whitehall east to beyond the Royal Exchange and from Bloomsbury south to the River accommodates some of them, but the most colossal roosts are usually those on St Paul's and the British Museum. Most of these birds do not spend the day in inner

London: they come in from outside, flying high in wisps and flocks and parties, and their movement is extraordinarily rapid so that, even on the chief flightlines, from the time when they begin passing a given point until the time the stream ceases only a very few minutes elapse. This presents serious difficulties in the way of observation. My first assumption was that these birds were winter visitors that spent the day feeding in flocks in the fields and open spaces of the outer ring of suburbs, and this apparently is the belief of most people who have ever troubled to make a guess at it. But I resolved to jump to no conclusion and to begin by following the main British Museum line out from the roost. I established it without much trouble coming in over Bloomsbury from the south-east corner of Regent's Park in a narrow and well-defined course. Above the Park the flocks were so large and compact in formation and were flying so high that I felt sure they must have come from a distance but no more observation was possible that day. This was the afternoon of Christmas Eve 1925: the weather was fine and most of the sky blue but one long ribbon of light gathering clouds ran across it from north-east to south-west and this the starlings followed as if it had been their road, flying both straight and fast.

It was the middle of January before an opportunity recurred to follow the line appreciably further. On this occasion I kept watch on the northern slope of Primrose Hill, which was then covered with snow, and my brother on the southern, so that together we commanded the widest possible view. Very soon the first small flock came in sight flying quite low over the hill and uncertain of their direction: they wheeled round it. The oncoming flocks grew larger and less hesitating. We struck up against the current across Adelaide Road and Fellows Road where the movement became vague and confusing, such large parties as there were being scattered and following no perceptible route, whilst twos and threes and single birds were extraordinarily numerous. Almost all were heading in the proper direction—south-east—but on a wide independent front and at no constant elevation. We now began to suspect that many had only just got up, and watching carefully we saw bird after bird rise from the gardens and chimney-pots on all sides and wing its way upwards to join the passing squadrons which consequently rose visibly in size as they went along. This indication that the flocks are not persistent but are formed afresh every afternoon out of aggregations of single birds was soon afterwards corroborated about Finsbury Park, where the birds congregate in the same way apparently to the St Paul's roost. This piece of observation practically shattered the common assumption that the great winter roosts of inner London are composed of starlings from abroad which spend the day feeding in the country. On the contrary, it is the ordinary suburban starling population, resident and stationary here, as ringing has proved it to be all over Britain, which prefers a communal roost in winter. The same fact was subsequently confirmed in the case of other suburbs, such as Barnes and Maida Vale and also in the central parks.[45]

It was particularly interesting to observe the behaviour of starlings in Kensington Gardens towards roosting time. In the late afternoon they mounted to the crowns of the trees and sang unceasingly to themselves or to each other. One by one they became impatient and took wing towards the Long Water, where on the fringe of the trees they used to halt to form loquacious gatherings in the upper branches. The babble of their conversation attracted many passing birds to descend and join them; by this process its volume rapidly swelled, casting an irresistible spell upon all the starlings within hearing. Nothing so fascinates a starling as the sound of other starlings: the gathering flock exercises over an extended radius a kind of magnetism or centripetal attraction, till all the solo performers have come into it. The largest of the assemblies was in the trees between *Peter Pan* and Temple Lodge: when it had attained a respectable size parties began to break off from it and fly over the bridge and down the Serpentine, the departures exceeded the fresh arrivals and the main flock, taking wing in a body, ascended towards Hyde Park Corner. Another day, watching in the same place, I saw one large flock come in from the north-east over the Long Water from the direction of Praed Street. This must have taken them far out of their direct

---

45. ED: In the chapter on Roosts and Fly-lines in *The Birds of the London Area since 1900* (1957), S. Cramp *et al.* wrote: 'Virtually no field work was carried out on the roosting of starlings in London until 1925, when E. M. Nicholson made a survey of the Inner London roosts and the fly-lines to them, the results of which have not been published. Up to this time it was commonly believed that the starlings came into London in a few enormous flocks and were probably Continental immigrants which had spent the day feeding at sewage farms and on playing fields. Nicholson showed that the roosting starlings were the resident birds of the suburbs and he described how the individual birds collected at a local rendezvous, then flew in towards London in small flocks, picking up others as they went.'

line to the central roosts. Though they concentrated in this way many dispersed again, leaving for the roost singly or in very small parties. All followed the north bank of the Serpentine.

In winter the whole number of starlings and woodpigeons and black-headed gulls in Kensington Gardens—the second, third and fourth most plentiful species, amounting to five sixths or thereabouts of the winter bird population of the Gardens, sparrows excluded—retire to roost elsewhere. Though there are approximately one thousand of them altogether, they form nothing but a drop in the ocean of the whole movement (not even forming the sole source of a single important flyline), which indicates how colossal it is.

Watching at Hyde Park Corner, inside the gates, the main starling highway from the west could be seen on the point of crossing the Bath and Exeter Road. The numbers at this point were spectacular: flock after flock came up without the least interruption and not one of them contained fewer than fifty birds—they ranged up to hundreds. The route continued across the Green Park into the Trafalgar Square roosts.

At only one other point were the observed numbers of starlings more impressive, and that was at the bridgehead by the Monument on the main route from the east to the St Paul's roost. It was a Sunday: a well-chosen day for the exploration of the City, its streets being empty of traffic and its pavements not swept, as they generally are, by irresistible currents of humanity. At Blackfriars we had seen small parties cross the River from the south—nevertheless it seems that the sleepers at the central roosts are drawn mostly from north of the Thames, and below Vauxhall the numbers that cross the River from south London seem never to be great—while others had been noticed crossing Queen Victoria Street towards St Paul's. Round this was a circling veil of starlings constantly flying so that it was almost impossible to be sure which were the newly arrived contingents and which had merely been flying round. Several satellite roosts on neighbouring buildings added to the confusion.

Farther east it was even worse. The streams flowed in varying and even in opposite directions. At the fairly large roosts on the Guildhall, and those on the Royal Exchange, St Michael's Cornhill and St Mary-le-Bow (some perhaps no more than halting-places), the main stream of birds came from the east but there were smaller cross-currents linking the roosts and traversing the general direction. We crossed a flightline in comparison with which these had been insignificant. At a very low estimate between three and five thousand birds must have passed in a very few minutes, only just clearing the gilded flames which crown the Monument. The numbers mentioned here and in other places may appear disappointing: it should be borne in mind that, without pretending to be exact, I have estimated numbers after actually counting a great many birds under varying circumstances and the figures are intended as figures, not as figures of speech. The usual tendency is to underestimate birds dispersed in cover and absurdly to overestimate the numbers of a massed flock: an otherwise careful observer will think nothing of speaking of tens of thousands when an actual count would probably have given twelve hundred at the most. The larger the bird the more exaggerated the impression: a hundred and fifty woodpigeons coming over to roost in a body will seem to darken the sky. If the convention of saying thousands when I have kept to hundreds and 'almost a million' where here you find only few beggarly thousands is necessary to your enjoyment, you may read them with a clear conscience, for they seemed at least as much as that. On the Monument many settled in the bristling tongues of the gilt flames and we found a roost in the flying spire of St Dunstan's in the East. They came from the direction of Tower Hill heading for St Paul's.

The results of other attempts to secure knowledge of the movements of the starling were not individually interesting and are best summed up together. One conspicuous circumstance was the extreme unevenness of the daytime distribution. They were not dispersed freely over all the suburbs but seemed to mass themselves in favoured parts so that it was possible to explore miles of streets without discovering any and then, coming into a district where the character of the houses changed, find it to be populous with starlings. One such district was Castlenau and the newer part of Barnes, others being about Clissold Park and Manor House, Swiss Cottage, Maida Vale and the parks. To the opposite sort belonged not only the slum areas but the more pretentious purlieus of South Kensington, Mayfair, Bayswater and Notting Hill: one afternoon I made a long and wearisome perambulation of the neighbourhood of Westbourne Grove and Ladbroke Grove and saw only one wretched starling warbling on a tree. They love the stock suburban type of house and garden; it would be roughly correct to say that, excluding slums, which are barren ground, the number of starlings on built-up areas varies inversely with the general height of the buildings. This correlation, such as it is, is due to the obvious fact that the height of the buildings is normally smallest when land values are low and most land is left free from them. The investigation led me to the conclusion that starling-London may in winter be divided into three clear zones: an outer one (including apparently Chingford and Higham Hill, Twickenham, Hampton and Sunbury, Kingston and, perhaps, Richmond) where they remain constantly, surrounding an inner zone where they are found by day but not by night and in the centre a little core or bull's-eye where they roost in great numbers but hardly any spend the day. There is hardly the slightest doubt that it would be possible to trace out the boundaries of these zones in detail. The boundary between the second and third appears to run from Charing Cross Bridge to the Admiralty, from there to Bloomsbury and on round the City to the Tower. Probably it varies from year to year, for the whole system is much too recent to be very stable. I have little personal knowledge of its history, but the inner London roosts, at any rate in their present form, seem to date from since the War. Mr Eric Parker in the *Spectator* gives 1919 as the year when perhaps six or eight starlings began to roost on the British Museum and states that twenty years ago there was none on St Paul's. But in St James's Park at any rate there was a roost long before.

There appears to have been a slow transition from the habit of roosting on trees and bushes, which is the usual practice in the country, to the London habit of roosting upon large buildings. At least I could not find in the central area a single considerable roost which was not on a building; the Savoy roost in the plane trees by the Chapel Royal was important while it lasted, but in 1925 was deserted after the first week of November. I was inclined at the time to put this down to something having given them a fright, but Mr Rudge Harding thought that they use this roost only as long as the large leaves of the planes offer them shelter and abandon it when those fall. A few roosted on islands in the park lakes, but the large flocks seen to alight in Battersea and Regent's Park always rose again eventually and continued. They love classical architecture and hate Gothic, not from any aesthetic considerations but because the horizontal edges give them shelter and accommodation, which the vertical do not.

It is probably for this reason that there is so far no roost at Westminster in spite of its great buildings. On the way in they pass over many opportunities of suitable sites without being tempted by them: I found only one small roost in the zone between the dormitory area and the outer districts where no movements take place. This was on Albert Bridge at Chelsea, where a few slept on the hideous iron towers; probably several others exist but I doubt if there are any of much importance. Starlings fly at forty miles an hour with ease and in London none can ever be much more than twenty minutes from the roost: spectacular as the daily migration seems its performance is far less wonderful than the organisation—it is hard to call it anything else—by which all these birds, spending their days scattered broadcast, not only roost together but reach the roosts by definite ascertainable lines of flight and already formed into flocks. Their convergence is carried out with marvellous speed and accuracy: by the time inner London is reached almost all are in compact bodies and the sight of an odd straggler comparatively rare. Wherever I have observed it concentration has followed the same simple plan, the single birds leaving at the proper time for a local rendezvous on the way to which they join others; at the rendezvous there is often a halt while others pour in, and the birds eventually continue their journey in a large body or in several bodies leaving one after the other. It is the same system as the collection of surplus rain into one large river by tributaries or sunlight transmitted into the trunk of a tree through leaves and branches. But the central roosts are not independent: overflow lines (the chief of which run south-east from Bloomsbury to St Paul's and both east and west between the City and St Paul's and Trafalgar Square) link them all together.

The movements of **woodpigeons** are less simple. There is concentration or dispersal based on a single centre and the operation is less perfect, for some do not take part in it. In the case of the gulls there is at sunset a general retirement to the suburban reservoirs, in that of the starlings a general movement towards the centre, but the woodpigeons journey instead to islands in the park lakes. Inner London has several roosts. The first is on the seven islets of Regent's Park lake: it attracts great numbers from the Park itself, from Hampstead Heath northward, from the King's Cross and Camden Town neighbourhoods and from the Bloomsbury Squares, at any rate down to Russell Square. The east flightline crossed the Euston Road between Tottenham Court Road and Unity House, the headquarters of the National Union of Railwaymen. They roost about Ken Wood also but undoubtedly a fair number from Hampstead Heath and Parliament Hill Fields go on to Regent's Park. The main part of the roost is on the islands near the end of the north-eastern arm. The Serpentine island roost is of less importance: the numbers present are often below a hundred—my brother watching on the shortest day found 83 in the trees by half-past three, and others on the ground. More streamed in, most coming from the north over the Hudson Memorial or the south east up the Serpentine, till the count reached 220. He estimated that there were 250 or more altogether, of which about 70, roughly, departed in bunches on the flightline over Chelsea: these figures were obtained by a count of arrivals and a check of the number of birds present. Mr A. Holte Macpherson records having counted over 400 here on 14th March 1924, but I have never seen such numbers.[46]

On recent occasions when I kept watch in Kensington Gardens it was clear that the great majority of the woodpigeons (except those immediately around the Long Water) ignored the Serpentine roost. Long before dusk they would ascend to the crowns of the trees and there sit placidly, alone or in little groups, so drowsy and hunched-up that they had the appearance of being already retired to roost. But their lethargy was brief and as the light began to fail they would become more restless, often changing their perch or taking short flights from tree to tree. Those from the north-west corner which were cantoned, as Gilbert White would have said, along the Broad Walk launched out in strings and parties over the Round Pond and alighted again on the far side, in the south-sloping part of the Gardens, where the chief concentration was.

---

46. ED: In 1897 Macpherson (*Nature Notes* 8: 46–47), noted that 'The island in the Serpentine was much used during the winter 1895–6 as a roosting-place for starlings and woodpigeons. The starlings used to arrive in hundreds shortly after sunset, flying in companies towards the island, usually from the west, until they got directly above it. They then dived down almost perpendicularly into the trees and shrubs. The woodpigeons, about fifty in number, had usually already taken up their position for the night before the arrival of the starlings. Both species continued to frequent the island throughout the summer, but in smaller numbers. Starlings seem fond of these metropolitan islands; they roost in the same way in Battersea and St. James' parks.'

From there, as dusk began to come up, they launched out in flocks over Gloucester Road and the Albert Hall steering a little east of south on the way to Battersea. From the Serpentine roost a considerable overflow flyline crossed the Rotten Row and Knightsbridge and Brompton Road, reaching the Thames about Chelsea Hospital and also ending at Battersea Park. From the east another converged on it: an important though broad and ill-defined line passing over Victoria and Ebury Street, which was used by the birds from St James's Park and the adjoining open spaces. Taken together these flightlines formed a half-open fan pivoted on Battersea Park lake. The roost there is very large and flourishing, gaining many more recruits from the south. Owing to the number and extent of the islands it is more scattered than the others and the lack of open surroundings to give a clear view also tends to make it less spectacular than the Regent's Park roost. The most I could count on one island was about 160.

The roost on Duck Island in St James's Park lake is of slight importance—it is possible that there is another small roost in Buckingham Palace grounds—and most of the birds from this part of London sleep at Battersea. Woodpigeons are also found sparingly at Gray's Inn and Lincoln's Inn Fields and other open spaces near the City, as well as on the Embankment, but owing to their insignificant numbers and unpunctual habits the unravelling of their roosting flightlines (if they have any fixed custom in this way) would require an effort out of all proportion to the knowledge that would be gained from it. I found, by the simple but tedious process of watching till it got too dark to see, that some of the birds which ascend to the treetops at Lincoln's Inn remained without stirring until nightfall, and presumably roosted on the spot. On the other hand, some from Lincoln's Inn Fields undoubtedly fly down over Kingsway and Trafalgar Square to either the St James's Park roost or the Battersea roost beyond, and it is not impossible that a few may join the considerable flyline from Russell Square, not far away, to Regent's Park. A bird by Smithfield Market started off at sunset north-eastwards in the direction of Victoria Park, where there was probably a roost, although I did not investigate it. Others from Clissold Park pointed in the same direction.

A good map of London with the open spaces coloured in green shows that, particularly between Regent's Park and Hyde Park, and between Hyde Park and the River, there are great numbers of little gardens and green places, occupying an unexpectedly high proportion of the whole area. Each of these as a rule has its woodpigeons: they even live in Soho. These seem usually to attach themselves not necessarily to the nearest roost but to the nearest line of flight—from Russell Square, for instance, they join the main stream from Bloomsbury further north, though the nearest roost is actually in St James's Park. From the Bayswater and Lancaster Gate squares they fly across to Kensington Gardens. In spite of confusing exceptions—I have seen three passing east along the Strand at roosting time and a large party striking off south from Regent's Park above Great Portland Street, possibly on an overflow line to Battersea—the main scheme of the woodpigeon movements was clear. Without moving so unanimously as the gulls they showed a very strong tendency to congregate to their moated strongholds, especially in the central area. Even at Walthamstow Reservoirs they were seen coming in to the impregnable island-roosts from all sides except the north. There seemed also to be a moated roost by the Mole at Esher Place, though this was strictly a country instance and may not have had the least connection with the metropolitan movements.

The woodpigeon, the starling and the common and black-headed gulls are the birds which make the longest and most regular of the daily migrations, and which travel in greatest numbers. But these are not all. The **carrion crow**, in spite of the tradition that he is solitary, resorts to a common roost. In this case the numbers are too small to constitute regular flylines, but I have seen a bird flying north over St James's and pairs over Regent's Park (twice) and also Haverstock Hill flying from there to Ken Wood, all going in the right direction about roosting time. That many must come in from all sides is clear from the size of the roost, for I have counted forty-two of them on the wing in a compact body. No birdlover who has the opportunity should miss hearing the music of these birds at the Ken Wood roost. Their cawing is deep and sonorous and each bird speaks with a different voice: heard in concert, as it rarely can be elsewhere, the effect is very fine. They were warmly sociable and when they were flushed all settled in a band in the crowns of two or three adjacent trees. But they did not all rise together: some—possibly the birds from the inner Parks—were quite trustful, the rest wild and shy. The flock as a whole was not easily approached. Walthamstow Reservoirs have another considerable roost. Some appeared to sleep in Hyde Park but I sometimes saw a pair after sunset flying rapidly west over Kensington Gardens: these undoubtedly went to some outside roost, perhaps no further away than Holland Park.

About midwinter **chaffinches** congregate in numbers at Ken Wood and apparently some come north all the way from Regent's Park to sleep there. The movements of the **house-sparrow** are petty and insignificant by comparison: nowhere in the central area did I find anything approaching a flyline. But the East End had one. About the East India Docks I noticed numbers of them flying overhead in the late afternoon—it was already the first of March—but all going in different directions and with no apparent order. But as roosting time drew near a very considerable flyline became apparent. I followed it back to the Iron Bridge, which crosses the Old Southend Road over the River Lea into Essex. There I found them still pouring over from East Ham in parties of up to forty or fifty birds, but more usually ten to twenty, flying silently about fifty feet up so that they scarcely cleared the higher buildings. At East India Dock a parallel stream became noticeable on the opposite—that is the south—side of the road, passing directly over the *Durham Castle* which lay there in her berth. Following them by East India Dock Road I came to their roost in the churchyard trees of All Saints, Poplar. One tree in particular, a plane standing alone in a children's playground just by Poplar Station, was completely thronged with them. The din was incessant and indescribable, by no means an unpleasant sound, not especially loud but shrill and far carrying. Both this and the way they fluttered restlessly and with extraordinary sprightliness about the twigs reminded me more than anything of the little exotic finches in the Small Birds' House at the Zoo. I attempted a count but it was impossible: they danced and fluttered and darted from place to place as unlike a sedentary species as birds could be—the noise and the crowd and the twilight seemed to go to their heads. They were all mad and very happy. The ground underneath was well whitewashed, proving that the roost was not a new one.

In such areas as the Docks where food—and sparrows—are plentiful but trees do not exist some sort of a movement at roosting time is inevitable and I have no doubt that it would be possible to discover many small local migrations. At Paddington Goods Yard there seems to be something of the same kind: the little groups of trees in central London are dormitories for all those sparrows in their neighbourhood and a certain amount of movement takes place between Kensington Gardens and Hyde Park.

I did not trace out the flylines of any other species, but there certainly were others. The **mallards** undoubtedly have a network of communications between the park lakes and the outer reservoirs: the most obvious runs from Barn Elms to the Round Pond and over *Physical Energy* to the Long Water, continuing down the Serpentine and over Hyde Park Corner to the lake in St James's Park. **Pochards** and **tufted ducks** use the same routes, but all these fly mostly by night: they have no roosting flylines in this sense.

It is impossible to tell what birds pass over London entirely at a height, without giving us any opportunity for observing them. It seems almost certain that the **heron** does so, for one. The great London heronries are in Richmond and Wanstead Parks:[47] take a direct line from Sidmouth Wood to Wanstead Ponds and it passes almost over Hyde Park Corner, leaving Barn Elms and the Serpentine so little to the left that any heron crossing from one to the other must see their water plainly. The comparatively rare occasions when one observes a heron low over London give little help: climbing or descending they are apt to circle very wide of their course. Last Christmas Eve I saw one from the top of a bus near Lancaster Gate: he flew quite low across the Bayswater Road and passed over Kensington Gardens to the Long Water. He came from the direction of Regent's Park. Another was seen flying along the Serpentine from the same end. Late on Easter Sunday 1925 a heron flew magisterially over the middle of Chelsea, going more or less west, towards Barn Elms, where I once made one cry suddenly in alarm when he caught sight of me standing a few feet below him as he flew in the dusk.

The **London pigeon**, the only abundant species in the central areas which has not yet been mentioned, shows much individual movement, but if there was marked order in it I could not find any.

---

47. ED: This is no longer true. Breeding appears to have ceased at Richmond Park in 1960 and the last nest at Wanstead Park was in 1957 (D. J. Montier 1977, *Atlas of Breeding Birds of the London Area*); however, herons were breeding again at Richmond Park in 1994, with seven pairs reported. Since 1968 herons have nested in central London in Regent's Park and, since 1990, in Battersea Park.

The **results of the inquiry** repaid the labour of making it. It became plain that, just as the metropolitan fauna with its newcoming gulls and woodpigeons was groping to adjust its composition to the abnormality of modern London, so the individual species were evolving new habits of daily travel to enable them to flourish under the new conditions. Some birds could live in central London which could not sleep there; others could roost there but not find food. With roosts, the main consideration appeared to be not comfort but safety. It was hard to see that the woodpigeons gained in comfort by congregating on trees on islands rather than on the mainland but they certainly gained in security. Cats might or might not be a menace to them but grey squirrels most certainly are: there is a case on record of one being overcome and eaten by the less powerful native red squirrel. The gulls retreat promptly to escape from all disturbance. The starlings found safety from London's innumerable cats on the high central buildings as they could never have hoped to find in their suburban haunts. Crows seemed to gather not so much from wisdom as for the sake of companionship. Only the house-sparrow dares to sleep on little trees: an arrangement of Nature apparently in the interests of owls and cats rather than the sparrows. Understandably some creatures protect themselves not by protective measures but by sheer invincible fecundity: they may be massacred perpetually and raise no obstacle, only they never diminish, that is the salvation of the species—the individual has no salvation.

The extreme lengths of the main flylines of the gulls to the Molesey roost were seventeen miles from the Pool of London, sixteen miles from Hampstead South End Ponds and thirteen miles from the Serpentine; the distances to the presumed Littleton roost would be at least two miles more. Only a single case was observed of a starling flight line which appeared to be more than five miles long—an unimportant line from Wimbledon apparently ending in the Trafalgar Square roosts, about seven miles away. Others of this length are probably not uncommon, but I find no reason to suppose that in the winter of 1925 any starlings were regularly performing journeys of ten miles or more. Their system is new and expanding: there is nothing to prevent a considerable extension in the future.[48]

---

48. ED: The early history of the starling roosts has been detailed by R. S. R. Fitter in *London's Natural History* (1945) and *London's Birds* (1949).

The woodpigeon's flylines were a little shorter. In time therefore no bird seemed to range even an hour's flight from its roost, though in distance it might be twenty. This much had been discovered by the time the investigation was concluded. A further inquiry would undoubtedly lead to the finding of many still uncharted flylines, especially in the eastern and southern suburbs. It might even bring to light something which would make it necessary for some of my conclusions to be modified, for I do not claim that they are all infallible. Undoubtedly the flylines change from year to year: after five years an observer going over the same ground would be sure to notice differences. All this adds to the fascination: the fact that the bird-life of London is changing infinitely faster than the bird-life of country places is the very reason for its peculiar interest to an ornithologist. The birds may be mostly commonplace; it is in their new reactions to us and to the metropolis and to one another that the chief value and fascination of bird-watching in London is to be found.

There was something I found profoundly depressing in the discovery, so consistently and relentlessly confirmed, that what one had imagined to be London—Piccadilly, the City, Oxford Street, Westminster—was no more than a little village embedded in the midst of an almost limitless dreary waste of streets and buildings more wretched in the appalling quantity than in the quality of their sordidness. One knows this of course from the map; one sees glimpses of it from the train; but no one can understand its utter hopelessness who has not undergone some such experience as wandering day after day through mile after mile of it scanning its murky sky for a redeeming bird. At this time, when the fascination of the quest was dying away, and my brother, whose help had been absolutely invaluable in unravelling mysteries and in sharing the labours of field observation had to go north, I decided to leave London. Early in March, just before the flylines broke down on the approach of summer, my inquiry was discontinued.[49]

---

49. ED: An enquiry by the Ornithological Section of the London Natural History Society in 1949–1952 showed that the starling roosts and flylines were by then also a feature of the summer months. The roosts reached their peak in July when young birds from the second and late broods joined them.

Chapter IV

Kensington Gardens and Hyde Park

The fauna of London changes as rapidly as London itself.[50] The mere spread of the houses has driven away birds which still flourished in the centre so long as there was country within reach of it.[51] The rook is one of these. Within living memory there was a famous rookery in Kensington Gardens; the old trees were wantonly chopped down and the rooks deserted, but in any case they would probably have had to go. The rook is in many places a town bird. The Promenade at Cheltenham boasts a fine rookery, or rather tolerates one.[52] I have seen rooks at the nest in Llangollen churchyard, and fourteen nests in a single tree above Godalming railway station; they reared young in the very heart of Cologne during the British Occupation. But though they flourish in towns they cannot exist without access to cultivated land, or some good substitute for it. On this account the inner London rookeries have vanished one after another and at the present time[53] the range of the rook about London penetrates very little way inside the five-mile radius from Charing Cross. It comes in to Wormwood Scrubs in the west, to Wimbledon Common and the Wandle in the south-west, to Brockley in the south-east and in the north down to the neighbourhood of Alexandra Palace. Precise boundaries probably do not exist, but inner London has been very thoroughly deserted: in my experience it has been easier to find a chiffchaff or a sedge-

---

50. ED: In 1910 A. H. Macpherson (*Selborne Magazine* 21: 65–67) considered the changes which had taken place during the previous twenty years. He wrote that 'The Serpentine, of course, presents quite a different appearance: it is well stocked with birds of many species which twenty years ago were complete strangers. Black-headed Gulls are now quite common throughout the winter: Wood Pigeons have become more numerous; and Starlings have also increased. On the other hand Rooks are now rarely seen and certainly fewer migrants rest on their way through London in the spring. But on the whole the birds have not changed very much. The [Carrion] Crow and Spotted Flycatcher manage to retain their position as nesting species, and Kensington Gardens is still as good a place as you can find for hearing the glorious song of the [Song] Thrush.'

51. ED: See also W. E. Glegg 1939, Changes of Bird-Life in relation to the Increase of London. *Lond. Bird Rep.* 3: 34–44.

52. ED: W. G. Teagle reports (*in litt.* 1994) that a few pairs were still nesting there in 1981.

53. ED: i.e. in 1926.

warbler or even a dunlin in the middle than it has been to see a rook there.[54]

Kensington Gardens were formerly rich in crows. Until seventy years ago magpies bred, and until about a hundred years ago ravens. There were two more which still remain, the jackdaw and the carrion crow. These are not merely the only surviving members of their family, they are practically the only surviving members of the old fauna that flourished when the place was a country park. Their contemporaries were nightingales and warblers and woodpeckers and red squirrels, birds and mammals which have long since disappeared. About fifty years ago only the remnants of this older fauna remained, and since then a new one has begun to take its place. The leading newcomers are the woodpigeon, which seems to have begun its great expansion during the eighteen seventies and eighties, the moorhen and the dabchick, which came about the same period, the black-headed gull, which was not regular or abundant until the hard winters of the eighteen nineties, the common gull which is still expanding now, the tufted duck which just bred on the Serpentine in 1924, and the semi-domesticated London pigeon. Survivors from the older fauna are the mistle-thrush, whose foothold in central London is not altogether secure, the blackbird, throstle,[55] dunnock, robin, chaffinch, and great and blue tits—all these seem to have diminished more or less seriously since the houses spread—the jackdaw, already mentioned, which has also gone down, the carrion crow, which seems to hold its own, and only three which have improved their position, the mallard (not of fully wild descent), the starling and the house-sparrow. Out of all these only the jackdaw and carrion crow, and perhaps the mistle-thrush, can

---

54. ED: In *Birds and Men* (1951) EMN wrote that 'although rooks like to feed quite close to the rookery they will cling to a favoured site even though all the land within three miles is sterilised by building. In Inner London rooks went on nesting in the city until a century ago, and the historic Inner Temple rookery was even recolonised without success as lately as 1916, a year after the Gray's Inn colony faded out. In Westminster and Kensington regular breeding ceased about sixty years ago, although sporadic attempts round Hyde Park continued into this century. In 1926 I noticed a still occupied small rookery near Earlsfield by the Portsmouth main line about 5 miles from Hyde Park Corner; going northwards at that time rooks became fairly plentiful from Alexandra Palace on. By 1945 only one rookery survived in the County of London, and there were only two others anywhere within 10 miles of St Pauls.'

55. ED: i.e. song thrush.

be called typical of the ancient fauna—the rest are common to most modern urban pleasure-grounds.

Some understanding of this is necessary to an understanding of the metropolitan bird-life, and especially of the bird-life of Kensington Gardens and Hyde Park.

When Yarrell[56] wrote at the outset of the Victorian age close on seventy species were said to have been noted in Kensington Gardens. During 1925 I myself observed sixty-one species there, and a more persistent observer might easily in this one year have eclipsed Yarrell's figure. But this is undoubtedly an indication not that the number of birds has not gone down but that the efficiency of observation has gone up. A much higher proportion of the modern figure has to be gleaned from occasional visitors. The following list shows the 64 species of wild birds observed by me in Hyde Park and Kensington Gardens during 1924–1926, together with their status so far as it could be ascertained. Unless the contrary is stated, the remarks apply to both the Park and the Gardens.

Birds of Kensington Gardens and Hyde Park seen in 1924–1926

Residents—16 species

Carrion crow Generally at least half a dozen present.

Jackdaw About four pairs near Kensington Palace.

Starling Common; they roost away from home in winter.

House-sparrow Exceedingly common.

Chaffinch Some remain but others are summer migrants.

Great tit Fairly common.

Blue tit More numerous than the last.

Throstle Not at all many, but the most persistent songsters.

Blackbird Chiefly about shrubberies.

Robin Also keeps to shrubberies, neither plentiful nor tame.

---

56. ED: W. Yarrell 1837–1843, *A History of British Birds*.

Dunnock Only a few, chiefly about shrubberies.

Mallard Plentiful at all seasons on the Serpentine and the Round Pond.

Gadwall A solitary drake has stayed now for several years.

Tufted duck Commonest as a bird of passage, but many winter and a pair or two now breed.

Moorhen Most by the Long Water, also the Serpentine, Dutch Garden[57] and in the Dell, but not the Round Pond.

Woodpigeon Plentiful; some immigration during the winter.

Doubtful residents—3 species

Mistle-thrush Bred 1925 in the Gardens, but apparently absent from mid June till October, and again rare in midwinter.

Greenfinch Often seen during May and apparently bred, also September, October and very often November; probably not a constant resident.

Pied wagtail Generally reckoned a winter visitor (September–April), but bred in 1925.

Summer migrant which breeds regularly—1 species

Spotted flycatcher By the Ranger's Lodge and elsewhere, but commonest on autumn passage.

Summer migrant which breeds occasionally—1 species[58]

Willow-wren[59] Commonest on autumn passage; used to breed, but not in 1924 or 1925.

---

57. ED: The Dutch Garden, which lies between the Orangery and Kensington Palace, is now known as the Sunken Garden.

58. EMN: The lesser whitethroat bred in Hyde Park in 1921 and is also seen on passage, but I have never definitely identified it myself.

59. ED: i.e. willow warbler; willow-wren was also used as a generic term including the two other *Phylloscopus* warblers, chiffchaff and wood warbler.

Kensington Gardens and Hyde Park

Summer migrants which seem not to breed—3 species

Swift Fairly plentiful in May and July, less so June and August, but some visible almost daily during their stay.

Chiffchaff Really a bird of passage, but in 1925 one continued in the Hudson Sanctuary, singing, at least till mid June.

Whitethroat Status as last, but prefers the Long Water Sanctuary.

Winter migrants—10 species[60]

Kestrel Not always present, but often seen November–March.

Black-headed gull October onwards; most leave in March but some linger until the end of May.

Common gull Much less plentiful than the last, appears later and leaves earlier.

Herring gull Only comes in hard weather, November–February.

Pochard On the Round Pond and the Serpentine; fairly numerous.

Wren A fair influx in October; some stay till early spring.

Fieldfare In hard weather, from November on.

Redwing Appears earlier and stays more often than the last.

Coot On the Serpentine and the Long Water, November onwards; not on the Round Pond.

Wigeon Only in severe cold.

Birds occasionally making a stay—6 species

Coal tit October and November.

Pied woodpecker[61] Observed only in Kensington Gardens.

Barred woodpecker[62] Ditto; March and April.

---

60. EMN: The tufted duck and pied wagtail, already mentioned, are principally winter visitors.
61. ED: i.e. great spotted woodpecker.
62. ED: i.e. lesser spotted woodpecker.

Cuckoo In 1925 one stayed from the end of June and through most of July, apparently.

Little owl First noted in Hyde Park 12th September 1925; probably a future resident.

Great crested grebe On the Round Pond and the Serpentine.

Birds of passage—16 species

Skylark April and October–November, also in hard weather.

Meadow-pipit April and September–November.

Yellow wagtail May and August.

Whinchat May.

Stonechat Long Water Sanctuary, October.

Garden warbler May.

Sedge-warbler May; sometimes remains by the Long Water.

Wood-wren[63] May and September.

Pied flycatcher August; probably accidental.

Swallow May and September–October.

House martin Observed only in September.

Sand martin May and September.

Turtle-dove May and August.

Lesser black-backed gull September.

Common sandpiper May.

Dunlin August; also observed in hard weather.

Occasional visitors—8 species

Twite November.

Linnet May.

Heron By the Serpentine and the Long Water.

---

63. ED: i.e. wood warbler.

Sparrowhawk October.

Grey wagtail October.

Lapwing October and November, in 1925, but not most years.

Stock dove December and January.

Brambling A finch, probably of this species, seen in December 1925.

This list is not based on nearly a long enough period of observation to be complete, if such things ever can be complete. It would be easy to make additions to the last two categories—for example the kingfisher, which I happened never to find here, and the marsh-warbler which a more fortunate observer once found singing by the Long Water.[64] My intention is simply to analyse the list of species in case it may be of use to other observers, or for the sake of comparison—I can vouch for it as representing pretty closely the position between March 1925 and March 1926.

In the country, on a comparable basis, one would expect to find twice as many undoubted residents and probably a dozen summer visitors, but there would be fewer winter visitors and far fewer which could strictly be described as birds of passage, and the total might be almost the same.

Undoubtedly the ducks and the gulls are the most perennially interesting to watch of the birds in Kensington Gardens and Hyde Park, so much so that to have included them here would have swollen this chapter to an unmanageable length and it was found

---

64. ED: Several other unusual species were recorded at about this time from the Gardens or the Park. References to these include: F. Newman 1920, Snipe in Kensington Gardens. *The Field* 136: 637; A. R. Severn 1921, Jays in Hyde Park. *The Field* 137: 657 (three near Hyde Park Corner on 8th May 1921); F. R. Roberts 1921, Nightjar in Hyde Park. *The Field* 137: 710 (one flying along the Serpentine near the bridge on 19th May 1921); H. Russell 1922, Tree-Creeper in Kensington Gardens. *The Field* 139: 66 (seen on 23rd December 1921); R. Clark 1922, A Partridge in Kensington Gardens. *The Field* 139: 172; F. Newman 1922, Woodcock in Kensington Gardens. *The Field* 139: 431; an adult male red-breasted merganser on the Serpentine on 12th February 1922 (*The Field*, 25th February 1922); A. H. Macpherson 1929, A List of the Birds of Inner London. *Brit. Birds* 22: 222–244 (several further species including a ring-ouzel seen in Kensington Gardens by Mr J. R. Harding on 28th April 1922 and several records of scaup on the Round Pond and the Serpentine). See also J. R. Harding 1922, Bird-Life in Kensington Gardens. *The Field* 139: 658; J. B. Watson 1922, Wood Wren in Kensington Gardens. *The Field* 139: 729; H. Russell 1922, Greenfinch in Hyde Park. *The Field* 140: 729.

necessary to grant each of them a chapter to itself.[65] But even with these sacrificed the Gardens and the Park are still rich in notable things.

I have already mentioned that Kensington Gardens and Hyde Park, being so completely isolated from other areas suitable for birds, except the most urban, form an uncommonly perfect gauge of the ebb and flow of migration. On any estate in the country one may observe from time to time stray birds which certainly do not belong there, but in the case of many interesting migrants, such as the skylark and meadow-pipit, the wanderers may as well be of the local resident stock as travellers from a distance. In central London that difficulty does not exist: when such a bird is seen it has undoubtedly travelled at any rate several miles, and it has not simply wandered from hedge to hedge or field to field, but embarked on a flight over unknown and barren country.

The migrations of the skylark and the **meadow-pipit** are particularly suitable for observation in the London parks. Neither species seems ever to make a long stay; it is not often that they even alight, although I have on different occasions put up both of them from the turf, especially near the Round Pond, where the migrating meadow-pipits are sometimes as tame as the sparrows.[66] But it was on the autumn passage that they were most conspicuous. The first were noticed in the second week of September. A month later others were still passing west in small numbers over the Round Pond, and I continued to observe them at intervals until 12th November, on which date one crossed eastwards.

The **skylark** passes over inner London in very much greater force. In the autumn of 1925 I first noticed this migration towards midnight on 16th October, when the flight-note, mixed with the calls of incoming redwings, could be heard over Bayswater. The next day I found others over Kensington Gardens (two of them settled on the grass near the Round Pond) and on the 22nd, 23rd and 24th the stream of larks flying at a hundred and fifty to three hundred feet westwards over the Round Pond was considerable and unbroken. On the 26th, after watching over three hundred skylarks, mostly in small flocks and straggling parties, travel west

---

65. ED: Chapters 1 and 2, respectively.

66. EMN: In 1925 this was very noticeable on 1st April.

over the Round Pond in the space of forty minutes, I noticed that the peak of the migration seemed not yet to have been passed, though it had now continued ten days. But after the 26th its force suddenly diminished, though all the remaining days of the month and for the first five of November odd birds continued to be seen. Then the migration seemed to cease, though a solitary wanderer was observed as late as 16th November. Mr Rudge Harding noted the passage over Richmond and on one occasion over Staines Reservoirs during the same period. But taking Kensington Gardens alone it seemed that at the lowest possible estimate fifty thousand birds must have passed over within the range of naked vision while the movement lasted.

Where had all these skylarks come from, and where were they all going to? The direction of the movement was consistently due west, and parallel to the Bath and Oxford Roads[67] on either side of it. This seemed to indicate that they came up the Thames Estuary (probably after crossing from the Continent) and followed the valley upstream. Some, perhaps the majority, probably halt in Berkshire and Buckinghamshire to winter, for when I was in that district I found great flocks in the fields (for instance on the Chiltern spurs between High Wycombe and the Thames) which were certainly migratory; these stayed at any rate till the end of February. Others perhaps pass right across the island, though it is difficult to guess what they do when they come to the sea again. Skylarks certainly appeared in large numbers at this time (23rd October in 1924) on the north coast of Devon, but some at least of these were seen arriving from the west, and had therefore probably crossed from South Wales. Most probably the main body settle down in the basin of the Thames. But immediately west of London the numbers in winter were, so far as I could find, insignificant. These daylight migrations of the skylark could probably be traced out if enough observers along the routes made careful notes of them.

An even more conspicuous example was the great trek southwards which took place on 15th January 1926—the skylarks' Retreat from Moscow. After three days of intense cold, with an east wind and sharp frost, snow fell more heavily than ever in London

---

67. ED: i.e. the A4 and the A40, respectively.

that morning, and all the country northward had locked itself against them like an oyster. This, and not the cold itself, is the real tragedy of hard weather—all their wide feeding-grounds close with a snap and they have to fly for their lives. Down at the Round Pond in the morning the movement was very conspicuous: they were heading due south at 150–250 feet, very silent and in flocks of a hundred birds or more. Later, looking out of my window, I found the movement in full swing just over the house-tops, and on such an impressive scale that I took a count. This showed that 315 larks passed in 15 minutes over a front about a hundred yards broad, all moving south except a few odd birds which cut across the line of flight westward. That would give 1,260 in an hour over this one spot, or more than 21,000 birds per hour over every mile of the front. I checked this rate at intervals and it did not fall until about four o'clock, when the stream stopped. About one o'clock I counted over 180 in three minutes; my figure was most likely an underestimate. To my knowledge the movement went on at this rate for more than six hours; how far it extended I have no information, but if the front were only eight miles broad it must have involved over a million birds. The mere figure makes little impression: we are accustomed to large figures and they have lost the power to make us wonder, even if they had not in the particular case of migration been so often fantastically exaggerated that no one pays much attention to them now. But to any observer with some knowledge of bird populations the number, for such a sudden and purely insular movement, may well appear unbelievable—many counties must have been emptied of their larks to recruit such an army. That the migration was on a considerable front was shown by the number of reports of it which appeared in the Press. At less than thirty miles an hour (and they were certainly covering more) the birds which passed over the Round Pond about ten o'clock would strike the Sussex coast, probably between Brighton and Beachy Head, by midday. From there they might have made the passage to Normandy by half-past two, or more probably turned west along the coast (about Worthing in a cold spell I often saw them following this route) until they reached some part comparatively free from snow. I heard that a great many settled down on the coast of Hampshire and ate everything that could be eaten there; fortunately the cold weather broke and the next day there were hardly any over London. When the worst comes to the worst they continue their journey to Land's End, where the risk of disaster turns almost to a certainty.[68]

The migrations of the **redwing** and the fieldfare through the central parks have something of the same character. They are, in the same way, visible and often diurnal, and they respond as closely to the weather. I first noticed the redwings at the same time as the skylarks, near midnight on 16th October, when the unmistakable flight-note sounded from passing flocks over Bayswater. The next day there were plenty in Kensington Gardens, wary and perching as they love to do on the crowns of lofty trees. All through that week the passage continued, and at night I often heard the whispered call. It went on intermittently till November, after which, the main immigration being complete, their appearances in the centre began to be regulated by the barometer. On 22nd November, in bitter cold, there were considerable flocks flying north-west up the Serpentine in the morning, at about 250 feet or less. During the severe frost at the opening of December they were again in evidence, and on the 5th we found one worm-hunting like a throstle on a freshly dug border by the southern Flower Walk of Kensington Gardens. A fortnight later I saw two or three feeding on the lawn where the rabbits are by the Dell in Hyde Park. They remained present though it thawed and turned to rain. I had never seen so many in Kensington Gardens as there were on Christmas Day. In the enclosure west of the Long Water nearly every evergreen held half a dozen. They used the loud clacking fieldfare-like note as well as the complaining one and I thought I heard snatches of the spring chorus, which English observers usually call the song: the true song is very different and not usually heard in this country. On 15th January—the day of the great passage of skylarks—many parties were seen pushing south-west, a few due south and others almost west. I noticed some in the Dell on the 28th, and on 5th March a spell of cold anticyclonic weather brought them back to Kensington Gardens, where I had seen none for some time. Two at least sang the low warbling chorus in the planes near Speke's Monument. I have dwelt upon the frequent appearances of the redwing in some detail because I have seen it assumed, in what should be a reliable quarter, that there is something exceptional,

---

68. ED: See also A. Gibbs & D. I. M. Wallace 1961, Four Million Birds? *Lond. Bird Rep.* 25: 61–68.

almost portentous, in the appearance of redwings in central London.

The **fieldfare** is decidedly scarcer. Until the census of 4th November[69] I saw none at all but on the 7th a considerable flock passed west over Kensington Palace Field, all silent. On the 25th, seeing a soft-coloured dove-like bird flying in front of trees north of the Round Pond, I put up the glasses and found it to be a fieldfare. It settled near the top of a low tree and, going over, I watched it softly chacking and chuckling to itself rather like a magpie. Either the sight or the strange sound attracted curious sparrows and soon there were thirteen of them crowding on the branches around and below it, displaying a close and silent but apparently unfriendly curiosity. The fieldfare flew off protesting and two more joined him; the soft French-blue of rumps and heads showed up well in the sun. On the morning of the 27th in a severe snowstorm I heard the characteristic cries and saw ten pass north-east above the Fountains at about 200 feet.

On 3rd December 1925, and again on 15th January 1926, I saw a party of **stock doves**, evidently frost-driven, in flight over Kensington Gardens. On the misty morning of 31st October 1925 and also on 29th November I noticed a solitary **lapwing** over the Round Pond; on the latter date I afterwards found two more at the lower end of the Serpentine. They were flying up and down at only about twenty feet, and apparently desperate for food, so that they ignored the crowds of people. But it was impossible to lure them to scramble among the gulls. Mr Harold Russell, who flushed one at the Round Pond during the same week, reported it as only the second in thirty years' watching; possibly the weather, exceptionally severe for November, was responsible for their occurrence. The scarcity of the lapwing in central London is hard to understand, for like the rook it has regular haunts at no great distance out.

The **dunlin**, which does not breed within two hundred miles,[70] might be expected to be much less seen, but I have found it on two

---

69. ED: 1925: see the next chapter.

70. ED: Dunlin now breed closer to London than this and non-breeding birds also summer on the east and south coasts of England (J. B. Reid in *The New Atlas of Breeding Birds in Britain and Ireland: 1988–1991* [D. W. Gibbons et al. 1993: 172–173]).

occasions. The first was on 21st August 1924, when as I walked round the northern end of the Hudson Sanctuary the peculiar call of the species caught my ear, and the dunlin, which was flying at about thirty feet above the ground, crossed the corner of the Sanctuary from the direction of Lancaster Gate and vanished travelling fast towards Hyde Park Corner. It was about midday. I saw no more in London till 16th December 1925, when I was surprised to find two of them flying about the still unfrozen part at the west side of the Round Pond at about 1.45 p.m. They settled on the asphalt edge and began to run towards me, eventually coming within ten yards. They were very tame, running along the margin like sandpipers and suffering a certain amount of persecution from the mallards and a few of the gulls, which would swim maliciously towards them and, coming on shore, drive them back a little way across the asphalt. Their small active forms, mouse-brown above and dull white below, with thick black feet and down-curved bills, looked unreal in such a setting. After some time they took wing and settled on the edge of the ice; then again to the asphalt, but the shore was barren and disturbers frequent, so that at last they flew off towards Kensington Palace, and actually settled on the middle of the Broad Walk, opposite the statue of Queen Victoria.[71] Flushed by some pedestrians they rose in the direction of the Orangery and disappeared.

On the morning of 21st November 1925, about which time an exceptionally dense and prolonged fog hung over the north Midlands, the weather at Kensington being more than hazy, I was standing by the Round Pond between ten and eleven o'clock when I suddenly heard the characteristic rather feeble call-note of the **twite** and, after looking round a little, saw two of them flying quite low across the Pond from the west and pass overhead with the usual rather nervous unballasted flight, disappearing among the trees towards Hyde Park.[72] Another winter finch which occurred about this time was a **brambling** by the Long Water on 3rd December. I have little doubt that it was a brambling, though the

---

71. ED: The statue is by her daughter H.R.H. Princess Louise, Marchioness of Lorne; it lies between Kensington Palace and the Round Pond.
72. ED: See E. M. Nicholson 1925, Twite in Kensington Gardens. *The Field* 146: 978. This is the first acceptable record for the London Area this century; the next three records each also involved two birds together.

circumstances made it impossible absolutely to swear to the identification. It shot up by the Sanctuary with a characteristic note which would have put the record beyond all doubt but for the fact that I had not met with the species for twenty months so that it was not sufficiently fresh in my memory.[73] The **linnet**, which differs from both in being a common resident in the south, is apparently not less rare in central London. I saw it only once for certain, in the Hudson Sanctuary in May; it is one of the species most likely to be encouraged to breed by suitable provision. The **grey wagtail**, not an uncommon winter visitor to the metropolis, I have noticed only once, by the Long Water in October.[74]

A **sparrowhawk**, perhaps from no farther than Ken Wood, which it still inhabits, was seen between the Round Pond and the Albert Memorial on 17th October. It flew past and settled in a tree near St Govor's Well.[75] The woodpigeon which had been sitting in the same place knew a sparrowhawk when he saw one and left in a hurry. Even the most trustful and stationary of London birds show no sign of forgetting their natural enemies. The panic caused by the dropping of a plank or a bucket, the back-firing of a petrol engine and similar unexpected reports amounts almost to shell-shock. The instant stampede shows how deeply fear of the gun has become ingrained in them. Examples are very plentiful. When the builders were active towards Kensington High Street their sudden commotions set all the sparrows by the Round Pond in a panic. One day in Trafalgar Square I saw three men who were nominally engaged in sweeping the slushy pavement stand waiting until all the pigeons had settled down and then suddenly kick over a bucket, provided by the Westminster Corporation for another purpose: instantly the flock sprang up in alarm and circled round Nelson's Column in close order. It was some time before they

---

73. ED: Four were seen soon afterwards, on 17th January 1926 during very cold weather, in Chelsea (A. H. Macpherson 1929, A List of the Birds of Inner London. Brit. Birds 22: 227).

74. ED: In *Birds and Men* (1951) EMN noted that 'Some not very frequent migrants and birds of passage, such as the grey wagtail, which pass over central London indiscriminately can more often be noted at points in the densely built-up area than other species such as the robin which are common not far away, but will not overlap their natural habitat.'

75. ED: St Govor's Well lies between the Broad Walk and the band-stand in Kensington Gardens.

settled down again, and then they did not return to the ground but alighted on the ledges. At the British Museum and elsewhere the stampede caused by a back-fire sometimes compels the bystanders to duck their heads, and in St James's Park the sudden insistent beating of a drum from the barracks across Birdcage Walk set all the woodpigeons rushing off in flocks as if they were being decimated by shot. Mallards are also susceptible and the salute of 41 guns at the opening of Parliament in 1926 made the pigeons behave like Gadarene swine. In the country most small birds utterly ignore the noise of a gun: it is only the species which suffer from it that shy at the sound. Once, at the Round Pond, I noticed that the sending up of a large white kite with a long tail, which bore no resemblance to a bird, had such a terrifying effect upon the gulls that all of them either left the Pond or retired to the far end. It is not necessary to suppose that they mistook it for a bird of prey: more likely the feeling of something large hanging ominously over their heads so played upon their nerves that mere mental discomfort would force them to decamp.

Most of the common summer migrants, even those which do not breed in the metropolitan area, occur on passage more or less regularly. Good examples of this are the **yellow wagtail** and the whinchat. I saw no yellow wagtails in 1925 until 16th May when, hearing the cries of some lost wanderers of that species over the Round Pond in the mist, I caught sight of two of them flying low overhead, circling in hesitation. On 26th August five passed very low above me, calling loudly, as I stood by the Long Water Sanctuary. They came from the direction of Hyde Park and, crossing the Long Water, disappeared westwards towards the Round Pond. Before breakfast on 12th May 1925 I watched a very tame hen **whinchat** in the Hudson Sanctuary. She flew from the bushes to the railings and then to little newly-planted trees in the open park, keeping silent all the time. At the same place there were simultaneously in view a pair of **whitethroats** (or at any rate a cock and a hen) and a singing **sedge-warbler**—a distinctly unusual gathering of birds to find in the middle of London. The sedge-warbler was musical and sprightly; he mounted singing to the top of a hawthorn and once explored the foliage of a birch, flying about in full view. On the windy, rainy morning of 27th May I thought I heard a snatch of sedge-warbler music by the Long Water, and stopped to listen. After a long wait, with an occasional drowned clamour of harsh grating notes which might almost have come from a distant mistle-thrush, the unmistakable babbling warble broke out close at hand, and some time afterwards I saw him fly from a hawthorn overhanging the water on the east bank to another clump where I had a good view of him. At the same spot a more fortunate observer[76] has listened to the rare and melodious **marsh-warbler**.

I have met a **garden warbler** singing in the elms on the east side of the Round Pond. This was on the day before the whinchat and sedge-warbler and the two whitethroats were seen in the Hudson Sanctuary—11th May. The same morning, walking in Kensington Gardens before breakfast, I heard the delicious trill of the **wood-wren**, and found him in a sycamore near *Peter Pan*. Later I met him again on the other side, in an elm, from which he flew with direct rapid flight to a birch and so by stages to the group of trees by Speke's Monument. It was now eight o'clock and there were many people about: seeing this as he came near the crossing of the paths he ceased to sing, and probably did not resume. The song was delightful in itself; it was also interesting as showing how migrants will sing without restraint where they can have no intention of taking up territory. Moreover I was able to see the wanderer at the moment of realising the true nature of Kensington Gardens. As London wakes up the number of people about, small in the early hours, gradually increases, and together with the reviving pandemonium in the streets has a disturbing effect upon the migrants which have come to rest within the central parks. Between seven and eleven o'clock, according to their temperaments, comes the moment when they find the place intolerable and pass on inconspicuously elsewhere. It is interesting and amusing to watch their restiveness increase with the increasing disturbance. On this account the early morning is incomparably the best time of day for finding uncommon birds in London.

The rarest bird I have ever seen in inner London[77] was an exception to this rule, for I found it in the middle of the morning, when the Park was full of people. It was a **pied flycatcher** in the

---

76. ED: H. G. Alexander saw and heard one singing in Kensington Gardens on 5th June 1924. See H. G. Alexander 1925, Marsh Warbler in London. *Brit. Birds* 18: 242.

77. ED: The twites (see page 85) were, in fact, much rarer than the pied flycatcher.

trees between the Putting Green and *Physical Energy*, on 26th August 1925—a little undistinguished bird, for it was in autumn plumage, most of the spring magnificence having moulted off by that season, leaving the breast a peculiar creamy tint, the colour of a faded letter.[78] This particular neighbourhood of the Gardens was also at that time the favourite haunt of the **spotted flycatchers**; one day—2nd September 1925—I found quite sixty there and not one in the rest of the Gardens nor in any of Hyde Park. Compared with them the pied bird was a darker more velvety-brown on the upper parts; it was also both smaller and slighter and its white wing-bar large and conspicuous, of a curiously ragged shape. It hawked in silence from the upper branches of the trees; I might very easily have missed it. In the north and west the pied flycatcher is moderately well known; in the south it is rarely seen even on passage. The spotted flycatcher, on the other hand, is the one summer migrant which breeds unfailingly in the central parks. It arrives late—in 1925 I saw none in London before 16th May—and the breeding strength in Hyde Park and Kensington Gardens together is only a very few pairs. At least one of these succeeded in rearing a brood in 1925: on 4th August I saw the young being fed on the high branch of an elm by the Ranger's Lodge. The young have none of the precision and elegance of the parents: they are unkempt little birds with short ragged tails, spotted above and below; they make an unceasing shrill affirmation of their perpetual hungriness. By the Hudson Sanctuary I have seen three or four together, constantly hawking from the backs of the chairs. One of these, a fledgling, dropped to the ground and began picking up stems of dry grass, which it let fall after carrying them aimlessly about for a minute or two. I have noticed similar behaviour in a fledgling mistle-thrush; in that case I put it down to imitation of the parents which would just have been building a new nest for the second brood—it was early May. But seeing it done by an innocent flycatcher in the last week of August suggests that it may be spontaneous—a first manifestation of the dawning nest-building instinct, just as some of the faint subsong at this season is undoubtedly the awakening in the young cocks of the impulse to sing.

---

78. ED: See E. M. Nicholson 1925, The Birds of Kensington Gardens. *The Nineteenth Century and After*: 927.

Close to the favourite haunt by the Putting Green there was at the beginning of September a flourishing and exceedingly populous wasps' nest, but though I watched for some time in the hope of seeing a flycatcher take a wasp even those which perched within a few yards persistently ignored them. I have heard of a pair of flycatchers which used to waylay the honeybees as they left the hive, and did great execution amongst them. At the beginning of September the thermometer passed 80 degrees,[79] but on the night of the 4th–5th it fell to 30° in London, and at Lympne in Kent there were 5 degrees of frost. The flycatchers, which rarely arrive before May and are not given to lingering for the last rose of summer, can rarely undergo such an ordeal; nevertheless they did not all leave. On the 6th I still found them plentiful; even on the 12th there were still at least six in the old haunt by the Putting Green, and on the 13th I saw two in the Long Water Sanctuary. These were the last.

On spring migration the three willow-wrens are not conspicuous. I had seen the wood-wren in May and a **chiffchaff** which had appeared while I was away at Selborne settled down in the plane-trees by the Ranger's Lodge and about the Hudson Sanctuary and continued singing well into June. This bird, as we proved beyond all doubt, was highly ventriloquial—so much so that it was not always possible to say with confidence even which side of us it was, and when it was actually visible it was necessary to put the field-glasses on it and see its bill open and close in exact time with the sound to feel convinced that the singer was not really in quite a different direction. It seemed most deceptive coming from the crowns of lofty planes, which it loved, but even low down in a poplar by the Hudson Sanctuary the sound gave the illusion of coming from much farther away. I have never met with another chiffchaff which had this mysterious power of deception.

Autumn migration is on a very different scale. Moreover it begins surprisingly early. Before my experience in Kensington Gardens I never suspected that by the first week in August the southward movement of **willow-wrens** was already considerable. But on 4th August 1925 I found the banks of the Long Water quite overrun with them—both adults and young of the year. They were undoubtedly on passage, for not a single pair had remained

---

79.   ED: Fahrenheit.

through the summer. They fed freely on the ground, and were almost silent. They were very tame and often approached within arm's length of me; one flitted about some angelica (or another plant of that kind) which was full of sparrows, and I expected them to mob it, but it showed more inclination to torment them, flying in pursuit of one or two when they left their posts.

On 12th September I found all three species—chiffchaff, wood-wren and willow-wren—present together in a group of trees behind the Magazine.[80] It was a very bellicose **wood-wren**; my attention was first drawn to him by the harsh cries of a starling which he was fiercely pursuing. His delicate pointed wings looked like a flycatcher's. When he swept up into a lime and settled in the lower branches I was able to watch him at very close quarters, as he moved gracefully along the twigs, amongst the fading leaves, continually making sudden little stabs with his slender pointed bill. He is a highly specialised creature: the long bill and mobile neck and the very long flexed feet enable him to pick off the black insects which infest the leaves without apparent effort. Each morsel of his food is in itself so minute that he must continually be gathering, as busy with his dagger bill as a scavenger with his pronged spit. He was restless and active but quite silent and kept to the lowest branches of the limes. I was still watching him when suddenly he caught sight of a sparrow fifty yards away or more, darted off in pursuit like a falcon unhooded and, overhauling it, forced it to take refuge in a bush. After that I saw him no more.

The **chiffchaff** was more plentiful than either the willow-wren or the wood-wren. I saw one on 4th August and another on 6th September; on the 12th one was singing pretty freely in the limes near the Magazine. On the morning of the 13th, which was misty with the sun trying to break through, I found an astonishing passage of chiffchaffs, in both the Park and the Gardens, but particularly in the Hudson Sanctuary. Even the song was pretty frequent, and sometimes as perfect as in spring. I saw two or three feeding on the tall umbelliferous plants in the Long Water Sanctuary. There must have been well over a hundred altogether. On the 10th they were again numerous in Hyde Park, less so in Kensington Gardens. The song was frequent and I heard the

---

80. ED: A group of buildings just north of the Serpentine which were once a military depot.

subsong, sometimes called the 'exhaust note'—a low husky soliloquy. Once I saw a scuffle, one bird chasing another full pelt among the trees; one of them uttered a few notes of the song, but whether it was pursuer or pursued could not be decided. Another note I heard was similar to the squeaking of the two whitethroats, but more mechanical and less like a vole or shrew: it sounded exactly as if a squeaky perambulator, pushed a few feet and then stopped, might have been the cause of it, but it undoubtedly came from the bird. On 17th October in the Long Water Sanctuary I again listened to the song of the chiffchaff, faint at first but becoming by degrees almost as loud and clear as in spring. I saw a bird soon afterwards and heard another across the water; there were certainly two and very likely more. This is a late date, especially since the redwings were already in. But I have met with it later still on the south coast, and in the west it sometimes winters.

Twice in 1925 I observed the **turtle-dove**, one on 31st May and again on 29th August. On 21st May, at ten o'clock in the morning, when boats were becoming plentiful, I saw a pair of **common sandpipers** fly up the Long Water to the Fountains. A moment later they returned to the bridge and began to mount in circles above it, the well-known call drifting faintly down as their flickering shapes grew smaller, till at last they towered so high that they were lost to sight. The sandpiper is apparently a regular passage migrant; it is mainly a northern and western species, but not exclusively, for broods are not infrequently brought off in at least one locality within forty miles of London.[81]

The **swallow** and both martins visit the Gardens on passage. Three times in May and once in August I noticed swallows; in September the passage was considerable, especially above the Hudson Sanctuary and the Serpentine. Usually they were travelling west at about 100–200 feet, but on the 12th most flew northwards as if it had been spring. On 22nd October I was surprised, on putting up the glasses at some skylarks above the Round Pond, to find two distant swallows in the field of view. It was about 9.30 a.m.; they were flying at about 100 feet and passed south over the Orangery and Kensington Palace, finally disappearing in the line of St Mary Abbot's spire. There was at least a symbolic interest in seeing the

---

81.   ED: Near Nuneham, on the Thames south of Oxford, from where there were breeding records in 1904, 1907 and 1922.

two armies of migration thus actually crossing one another, for the skylarks and long skeins of chaffinches pouring consistently westward must have come with the redwings from the Continent, while the swallows were part of the rearguard of the summer hordes. On 27th October I saw two more swallows at almost exactly the same place. This is not my latest date for the swallow even in the London district, for in 1915 I noted it at Golders Hill, the northern part of Hampstead Heath, as late as the 29th. Every year about the middle of October the season for reporting 'late swallows' opens in the Press; but there are always a good many stragglers till the end of the month, and in Sussex and Devon I have seen them in November. **Sand martins** on passage were more plentiful than either house martins or swallows; as late as half-past ten on the morning of Sunday 10th May 1925 I saw two appear among swifts playing over the Round Pond and pass just above the heads of a great crowd of people before vanishing beyond Kensington Palace towards Notting Hill. On the 12th there were five of them flycatching over the Round Pond; the dates suggest how prolonged their passage is, for by the 13th I have found hard-set eggs by the River Lune in Westmorland, hundreds of miles farther north. Very few appeared on September passage, and I saw **house martins** only on the 12th.

Between the birds of passage proper and the resident species is a vague and rather difficult group of birds which were found to make a longer stay at various seasons but never permanently to settle down. The **stonechat** might be included among the birds of passage for, though it is not a full migrant, autumn movements are certainly considerable. In 1925 I found a hen or young bird of the year in the Long Water Sanctuary, very brisk and active. It did a good deal of flycatching and perched a little on the railings; once I saw it drop down onto the path and hop under a seat to retrieve something edible. Intermittently a sparrow or two persecuted it, causing it to fly off with a loud grating alarm; otherwise it was silent. I saw it on 17th October, and again on the 25th. Also on the 17th a **coal tit** appeared in Kensington Gardens among a flock of blues and greats; on the 24th there were numbers, and the musical spring call was freely uttered. They kept to the hollies and low evergreens more than the other tits. But either they were only passage migrants or the cold weather proved too much for them, since after the first week in November I saw no others.

About the same time as the coal tits came I first met with a much more interesting visitor. On 28th October, walking by the Long Water Sanctuary, I heard rather faintly the characteristic 'Tcheck' of the **pied woodpecker**, and my brother saw the bird a little later fly over to a group of willows half-way along the edge. Going over we were amazed to find the woodpecker—an adult female—busy on a half-excavated hollow, which was exactly like, even if it was not designed to be, a typical nesting chamber. Much progress had already been made, for the circular hole had been drilled to that awkward stage when the hewing of the downward shaft has to be begun, and the bird, still compelled to work from outside because of the lack of space within, had to crane so far that at every stab the tail jerked up and the rest of her form vanished in the recess, except for the large and striking crimson patch on her belly. Every now and again she broke off her task and climbed a couple of feet to the point where the trunk—which was only about a foot at the most in diameter—had been lopped clean off. Here she would sit prominently for a moment before flying to one of the neighbouring willows, down which she generally made her way before crossing to the base of her own tree and working up again to the hole. When she sat on the watch-tower or when she flew up to the crown of a willow she would utter a single 'Tcheck', not very loud, apparently as a signal, so that I thought she might have a mate near. In the end she flew over the water and did not come back as long as we waited, though we heard her cry once in the distance. The occurrence was doubly interesting, first as being the only bird of the species I had yet seen in Kensington Gardens, but far more because of the abnormal season for nest-hewing.

The next morning, returning to the spot, I found the construction had most wonderfully progressed, and the woodpecker, who was still hard at work, was now quite hidden except when she came up to eject chips or to have a look round. Before ejecting them she generally put her head cautiously out to make sure that the coast was clear; then, gathering up the loose chips, jerked them briskly out so that they scattered and fell to the ground. Sometimes only seven or eight beakfuls were disposed of at a time, sometimes as many as sixteen. Every now and again she left the hole as before and flew up to a commanding height on some tree, sometimes against the trunk, sometimes among the outer branches. Occasionally she uttered the 'Tcheck' note and was once or twice feebly pursued by a sparrow. When she returned from these excursions, which were always brief, she flew straight to the hole without bringing anything back; they seemed to be simply opportunities to stretch herself. The outside entrance was now considerably neater and a more perfect circle than on the preceding day.

On the morning of 30th October, when I came on the scene about 9.30, the woodpecker was tinkering with the trunk outside the nest-hole, tapping apparently with the idea of sounding the wood, for she did not attempt to hack any away. The hole was now evidently deep though still unfinished, but on coming up with chips she simply looked round and began to eject them by the beakful, scarcely disappearing from sight all the time. It is obvious that the chips must come from the bottom of the shaft, and also that the bird could not possibly have stooped down, as it were, and picked up every beakful in the very short intervals between each ejection; actually she was undoubtedly drawing her bill in and picking up the next load of chips from somewhere just in front of her upper breast. Apparently, therefore, she must collect all the rubbish she can from the bottom and form her breast, tail, and probably wings into a sort of apron against the wall of the nest, so that she can bring it all up at once and bale it out by the beakful. Today there were generally from eight to twelve of these at a time, but many were very meagre. Sparrows were beginning to hang suspiciously about the scene of operations: once when she emerged she was closely pursued, and another time an inquisitive cock-sparrow loitered about so near the entrance, trying to peer in, that she had to rush out and scare him away. An uproar followed behind the bushes: it seemed that at least a dozen other sparrows had come to his rescue, but the woodpecker retired in good order, and the sparrows, though they still hung about, became more circumspect.

The next day I watched some time without the woodpecker putting in an appearance, but on 1st November I found her at home again. The slight friction perceptible two days ago had now developed into a state of siege. Truculent starlings were on guard almost continually, a pair and often more of them, occasionally singing a snatch, and sparrows under the influence of a hostile curiosity loitered about near the entrance. The effect was curiously human: it reminded me rather of a house where a crime has been committed, with its guards and 'morbid curiosity seekers'. The woodpecker actively resented this impertinence, but could not put an end to it. At frequent intervals she would come to the mouth, glance out and see the intolerable starlings, make a sudden furious sortie before which all the tormentors scattered and retired in confusion, and follow one which she had marked down, sometimes chasing it in circles in and out among the branches, but never with any decisive effect. They did not give her time to regain her stronghold before they were all down again within a foot of it. Sometimes, seeing them returned, she would make a second attack with the same lack of effect, sometimes she bowed to the inevitable and went indoors again. Only once did she succeed in engaging the enemy—a starling—in close combat. We could not distinctly see what happened, for it was very quickly over, but both birds sank to the ground with their beating wings clashing and almost locked, and it looked as if the woodpecker was striking with her claws in an effort to rip the starling's breast. The effect of this demonstration on the others was only temporary and soon they became as troublesome as ever. They did not want her nest: when she at last flew right across to some trees near the Dogs' Cemetery[82] none attempted to enter, though while she was present they had almost poked their heads inside. They simply hated and distrusted and wanted to persecute her.

During these alarms and excursions I was amazed at the brilliance of the woodpecker. Against the trunk she was a rather dull black and white bird with an almost dirty buffish breast, but in flight she displayed such a bewildering variety of contrasting mirrors and patterns, and such a crimson glow, that when she was seen—as the bird rarely is in normal circumstances—flying before a good background close at hand and nearly on a level with the eyes she became truly brilliant. The flight, though neither quick nor especially graceful, has dignity and decision: once I saw her glide most beautifully down towards the nest with her frayed wings held motionless, raised on either side of the body. She threaded her way adroitly between the branches, and showed herself active in the chase and capable of very sudden swerves; my brother believed she executed an 'about turn' by bending the inside wing sharply at the

---

82. ED: The Dogs' Cemetery is by Victoria Gate in Hyde Park.

joint and pivoting round on it. This day the call was not once uttered: perhaps she had now realised that she was quite alone.

Besides the starlings and sparrows a robin took part in the woodpecker-baiting, which was also done in silence. When she was looking out she rested her chin and beak on the threshold; in order to look upwards it was necessary to turn the head on one side. Sometimes after making a clearance she would sit defiantly on top of a stump. I noted on this occasion that the persecution she was suffering threatened to drive her from the nest, if not from the Gardens, and from that day until 3rd December I never saw her at all. Then during the second census[83] she flew across the meadow east of the Long Water and I followed her on several short flights, sometimes attended by sparrows, till I lost touch very close to the hole. I never saw her again. The hole, which faces east-north-east, is plainly visible from the path about the keeper's sentry-box; it will probably remain as her monument for many years.

The **barred woodpecker** appears to visit Kensington Gardens fairly often; I met with it myself during March and early April, and watched one utter the feeble repeated cry from a low branch near *Physical Energy* on the morning of Easter Monday 1925.

A more conspicuous visitant, who embarrassed himself by appearing on an equally public occasion, was a **great crested grebe**, present on the Round Pond as late as ten o'clock on a fine Sunday morning in May. Coming down to the edge the stiff T-square of his prow caught my eye well out in the mist. After a few minutes people began to pour in, especially three or four men with colossal model yachts, which they launched and set racing across the Pond. Already the grebe had grown apprehensive and now it took flight—a large and grotesquely attenuated wild duck-shaped bird, with the appearance of a long solid tail (formed of the lobed leaden feet turned slightly upwards, at an angle to the body's axis, giving a curious buckled-up impression, like an injured dragonfly). The wings, set unnaturally far back, were fairly long but absurdly narrow, like flippers; their beats were fast yet deep and never clipped, and the pace in full flight was astonishing—I put down 60–70 miles an hour as a conservative estimate. He circled round fairly low and at last headed for the Serpentine, which the rowing

---

83. ED: See the next chapter.

must have made intolerable for he soon returned, spreading the broad flat feet in readiness while still thirty feet up, and dropping in an almost horizontal pose, without a trace of the oblique checking throw-up of the mallard, so that there was only the brake-power of the feet to bring him to a stop. Soon afterwards he rose again, without much splashing or difficulty but taking a long run westward; he mounted to 20–30 feet, swung into the Broad Walk and flew down it to about St Govor's Well, then turned over the Albert Memorial and disappeared above South Kensington, flying at only about a hundred feet. In flight the head and neck were exceedingly attenuated, tapering to the spear-like bill; both fore and aft of the wings the length was much greater than a mallard's. I mention these details because a flying grebe, although by no means rare, is an unfamiliar sight to many who know the swimming bird well. On 30th March 1925 I saw one on the Round Pond as late as noon.

The appearances of the **heron** and the **lesser black-backed gull** are mentioned in other chapters and need not here be repeated.[84] The **little owl** seems to have appeared in Hyde Park for the first time in the autumn of 1925[85] and when crossing the Park about midnight on 12th September I listened to the characteristic hullabaloo, sometimes very close at hand, though the bird itself was invisible. But most of my notes of it for inner London refer to the Bayswater district. The **tawny owl** I have contrived to miss seeing, though it occurs pretty freely and has been suspected to be a permanent resident in Kensington Gardens, which are invariably closed at dusk. Major Pollard tells me that while visiting the Broadcasting Station on the roof of Selfridges he has seen a tawny owl emerge from the darkness of Oxford Street into the full glare of the floodlighting.

The **cuckoo** I first observed here on 25th June 1925. Seeing a strange-looking bird flying at some distance over the Serpentine as I stood on the bridge, I put up the glasses and was surprised to find it a cuckoo. It crossed the water and passed quite close, flying west

---

84. ED: The chapter describing the occurrences of the heron is not extant—see the footnote on page vii; those of the lesser blackback are described in Chapter 2.

85. ED: See R. W. Hayman 1925, Little Owl at Kensington Gardens. *The Field* 146: 1021; one had previously been reported seen sitting on an elm in Kensington Gardens on 22nd April 1922 (J. B. Watson 1922, *The Field*, 6th May).

at about forty feet. The time was half-past five in the afternoon and the Park was crowded with people, for it was a fine sunny day.[86] A cuckoo was also heard calling at various dates subsequently by officials of the Natural History Museum, who reported it in *The Times*;[87] and on 30th July, as I passed near the Putting Green, I saw a cuckoo fly silently over towards Prince's Gate. There is more than a possibility that all these records refer to the same bird, which in that case must have stayed in the Gardens not less than five weeks.

Two other summer migrants which do not breed remained present for considerable periods. The **swift** is so constantly seen between the first week in May and the beginning of August as to make the suspicion that it breeds in central London almost irresistible. It might well do so and be far harder to detect than any little plover on his shingle wilderness, but the swift is a far-ranging bird, and the manner in which the numbers leaped up from one to thirty or more pairs on the approach of stormy weather led me gradually to believe that these Serpentine birds might easily breed as far off as Hampstead or Barnes.[88] It is possible that some are non-breeders.

A few days after seeing the pair of **whitethroats** on passage in the Hudson Sanctuary (12th May 1925) I found a cock singing by the Long Water, and he remained there throughout the second half of May. His brief but uncommonly melodious song could be heard quite plainly on the farther side of the Long Water above the roar of the traffic. He sang from the trees more often than in the undergrowth, once from the very top of a sixty-foot elm. He also ranged through the Hudson Sanctuary, and I found him singing close to *Rima* two days after she was unveiled.[89] His head, I noticed, was very brown, almost the same shade as the back; this is normally the sign of winter plumage. On the 21st he aroused false hopes by appearing with a stem of dry grass in his bill; the action I

---

86. ED: See E. M. Nicholson 1925, A Hyde Park Cuckoo. *The Field* 146: 182.

87. ED: Mr W. D. Lang heard one calling in Kensington Gardens on 14th and 15th July (*Times*, 16th July 1925).

88. ED: Dr Stuart Smith told Stanley Cramp that in June and July 1928 he had seen three pairs of swifts feeding young in nests under the eaves of the General Post Office headquarters in the City (S. Cramp & W. G. Teagle 1952, The Birds of Inner London, 1900–1950. *Brit. Birds* 45: 433–456).

89. ED: See footnote on page 110.

am sure was merely symbolic, the unconscious expression of a disappointed hope, and so far as I could find he had neither mate nor nest. He uttered the 'Wick-wick-wick' alarm-note freely, and the twinkling squeaking shrew-like call which the lesser whitethroat also has.

Of winter visitors the most important are dealt with elsewhere, in the chapters on ducks and gulls. The **coot** appears when winter has definitely set in and remains, generally in small numbers, till early spring; most haunt the part of the Serpentine south of the island, but some also explore the Long Water. Neither Mr Harold Russell nor I have ever noticed any on the Round Pond. Generally they are tame and a little cloddish. I have seen one make a curious attack on a black-headed gull sitting on one of the posts round the island, pecking viciously at its feet (by craning the neck and almost standing in the water) and so causing it to decamp. After this it departed, without showing any interest in the vacated post.

From October onwards I fairly frequently saw a **kestrel** about. Once, when I put up the glasses to look at a tit in the crown of a tree, I found him in the field of view: he glided high above the Fountains and a moment later began to descend with a magnificent steep plunge, extending the feet and half closing the wings till he attained a breathless speed. The descent resembled a stoop, but probably had no predatory aim. One seen on 5th March hovered at an unusual height and above dense trees.

On 17th October I noticed a **wren** for the first time and by the 25th the Gardens had become full of them. It was interesting to observe that though the weather was fine and even the chaffinch, as well as the robin and starling, was heard singing about this time the wrens did not sing at all. Where they are permanent residents they sing persistently in autumn; this scrap of negative evidence supports the territorial interpretation of bird-song.

The status of the **pied wagtail** is dubious. In winter it is often seen, and during the first week of January I have flushed one in the rain from a lime-tree well out in the middle of Kensington Gardens. I should have called it a winter visitor but for the fact that on 15th June 1925 I came across a cock feeding a family so young (being provided with only the most rudimentary stumps of tails) that they can hardly have been hatched far away. Two were sitting in the trees, calling incessantly, and two more on the ground; as long as I watched the old hen, if there was one surviving, failed to put in an appearance. It seemed curious and almost inexplicable that, except on this one day, I should not have seen them between April and September, though the pied wagtail is one of those flaunting birds—the jay is another—which is capable of making itself miraculously scarce at times. The whole occurrence is a mystery to me; the pied wagtail certainly is not known to breed regularly. Much the same applies to the **greenfinch**. I saw it too often in the breeding season not to suspect it of having a nest, especially since it frequented a favourable shrubbery near the Albert Memorial and was usually in pairs, but there was no actual proof, and except in autumn it was not often noted at other seasons. The **mistle-thrush** undoubtedly bred: on 16th May I saw a fledgling with his parents near Speke's Monument. Apparently it breeds most years, and later I saw five together—four adults and one young—in Kensington Palace Field. But from June until the beginning of October I saw none at all, and in January and February they were not much in evidence.[90]

---

90. ED: The following extract from EMN's Birds and Men (1951) is also of interest: 'Mistle-thrushes have few obvious enemies, and are well able to look after themselves. No bird preserves territory more jealously, or is more vigilant and stout-hearted in driving out predators known or suspected, including hawks, owls, crows, magpies, jays and cats. The grating "rough music" with which the mistle-thrush accompanies its assaults on such intruders is one of the most familiar bird sounds of spring in mistle-thrush country. The boldest mistle-thrush I ever knew built a large and conspicuous nest, visible fully 150 yards away, in the Green Park in 1947. It was in a fork less than 6 ft. up in a little tree within a yard of a busy path and within 100 yards of Buckingham Palace and one of the busiest traffic centres in London. There were four young, but it was impossible to count them unless the bird was away as it merely crouched on approach and when I actually touched it, pecked my hand. On one occasion I found near the nest an astonished man with a largish brown and white dog on a lead. As he moved near the nest one of the mistle-thrushes, entirely ignoring his presence, swooped repeatedly down at the dog from a height of some 20 ft. to within a foot of the dog's nose, the dog (still held on the lead) snapping back without effect. This happened about ten times in quick succession. I spoke to the man, who was under the impression that the bird had taken some unaccountable dislike to his dog and was quite unaware, like almost all who passed that way, of the nest just in front of him. The pair were fearless in taking food to the nest even when people were near. They foraged quite close, largely at any rate on the open grass. Unfortunately, shortly before they should have fledged, the young were found lying dead under the nest at the end of April, probably through human action. ... An interesting possible clue to the mistle-thrush's tastes is that in London it distinctly prefers the waterless Green Park, which is the least attractive to most species. The Green Park is more sloping than the others and contains groves of low hawthorn trees and one large clump of mountain ash, which is the favourite centre for the mistle-thrushes.'

The mistle-thrush is the largest and loudest songster in the Park, as it is in all England, but the **throstle** is by far the most persistent. As the census[91] showed, he was far from plentiful, but he is the special songster of the metropolis and his bawled minstrelry is the only kind which asserts itself with perfect success above the continual roar of the traffic. He repeats his notes for emphasis, like a heckled politician, defying interruption, and he is also, like the cuckoo, a bird who lets himself be heard to the uttermost unit of his numerical strength. Six throstles will produce a greater volume of sound in the season than a couple of dozen mistle-thrushes, notwithstanding that the mistle-thrush is a larger bird, and when he sings at all sings even louder. Close at hand, in the silence of the country, the throstle's music is intolerable, but the distant roar takes its harshness from it or, more likely, cures our ears of such fastidious perceptions and his voice sounds pure and gracious above the metropolitan din. **Blackbirds** are more plentiful but begin singing late; I have seen one fly warbling to the tree which overhangs the Hudson Memorial. **Robins** and **dunnocks** warble almost unheeded, their feeble voices annihilated in the sound-burdened atmosphere; both are scarce and keep to the shrubberies, though I have seen a robin in a snowstorm well out in the middle of Kensington Gardens, near Speke's Monument. The **chaffinch** is more easily heard, though not particularly common. In Kensington Gardens I did not hear the song in 1926 before the last week of February, but the year before it continued unusually late, and there was still one in full song on 2nd July. On the warm spring-like morning of 24th October I heard it three times in the Flower Walk near the Albert Memorial, rather feebly at first but the third time clear and ringing. In mid October I have seen stranger chaffinches passing westward across the park in flocks of from twenty to fifty birds. The migration lasted about a week.

The chaffinch undoubtedly belongs to that curious group of birds which are tamer in many parts of the country than they are in London. 'Pink-Pink', a wild chaffinch at Temple in Berkshire is arrogantly tame: he comes to the gardener's punctually for meals and, if ever the clock is slow and the meal is a little late, his complaints are voluble and hard to appease. He sang from the ground at my feet. At the Strid Cottage near Bolton Abbey there is a

---

91.   ED: See the next chapter.

cock chaffinch who comes freely into the room and under the table to be fed, but such confidence is very widespread except, it appears, in London. The Kensington Gardens chaffinches do not flaunt themselves and are not approachable: even in a thick fog and frost we have had serious difficulty in luring a hen to our crumbs, although there were few sparrows on the scene. The robin is undoubtedly less bold than in the country; in the central parks he is almost shy. Blackbirds, throstles and dunnocks are perhaps no less trustful than elsewhere, but mistle-thrushes certainly are, and to see them at close quarters one must go farther afield.

The **house-sparrow**, which outnumbers all others by two to one, is dealt with in other chapters,[92] and the **great and blue tits**, which both flourish fairly well, call for no comment, except that to meet with a country bird after growing used to the dingy London ones gives as sudden a shock as if you had never seen any before. Numbers of **starlings** breed in the Park and Gardens, which about the end of May become full of the harsh-voiced fledglings. They ignore mankind absolutely, although a few occasionally take some share of gifts of food: normally they forage on the trodden grass. The black cluster of birds all tumbling over one another with eagerness and giving tongue to a hurried but not unmusical doxology where the hunting proves good is like nothing else. The

---

92. ED: These are not extant—see the footnote on page vii—but in *Birds and Men* (1951) EMN recounted how 'A St James's Square house-sparrow not long ago formed the habit of flying into the main conference hall at the Ministry of Labour and going to roost on a ledge above the Minister's head during important meetings.' He continued as follows: 'With sparrows it is the voice that is most characteristic rather than particular call-notes. The cries are innumerable and often almost indescribable, but they rarely deceive or confuse. Some may be peculiar to the London birds; at least I have heard various calls in London that I never hear in the country. One of these is a feeble less abrupt echo of the pied woodpecker's "tcheck". The full song of the sparrow begins about the first week in February, a little before the chaffinch's, and soon becomes the most widespread attempt at metropolitan bird-music. I have listened to it in New Oxford Street, at the Bank of England, in Seven Sisters Road, on Buckingham Palace, on the chairs in Kensington Gardens, in Berkeley Square, on Hampstead Heath and in many other places. Musical notes are sometimes included, but these are simply aberrations; the normal song is a continuous soliloquy of rather spluttering excited chirps, a shrill jerky flow of notes, none particularly melodious but a few quite passable. It most resembles those of the tree-sparrow and the reed-warbler. Sometimes the singer cocks his tail in an attitude of display, but the performance is apparently almost, if not quite, unconnected with the holding of territory, and this no doubt is why so tough a bird sings less loudly than the wren or even perhaps than the goldcrest. It is in fact nearer a subsong than a true song, and is accordingly delivered without any attempt to select a prominent song-post.'

song, such as it is, goes on in season and out of season, but especially in warm rainy weather, and the starling is from August until midwinter the only considerable songster, for volume of sound, to be heard in inner London. We curse his exasperating cheerfulness when, for months on end, there is no other bird to listen to, but without his presence the metropolitan autumn would certainly be a sepulchral season for bird-music. His nesting behaviour is notoriously careless: unwanted blue eggs are laid and abandoned on the lawn, and though most of the successful broods are reared in May there is no invariable season. In the Long Water Sanctuary I have watched pairing in November, and this couple was then in occupation of a suitable nesting-box. Nest-building begins by the first week in March, before the winter roosts are abandoned.

The **jackdaw** is probably the least known of the sixteen resident species. Although his foraging expeditions take him far over Kensington and Bayswater, and he seems sometimes to stay away for days at a stretch, his usual haunt is limited to the south-western corner of the Gardens, west of the Broad Walk, between Kensington Palace and Kensington Gore. Here during 1925-26 the full strength amounted to four pairs; even this meagre remnant was a marked improvement on the years before. They are silent and almost secretive birds, very different from the inhabitants of larger colonies. Occasionally one or two of them will throw this restraint to the winds and set up such a din that the impression is of an invading flock. In April I found a pair building in the hollow top of an elm in the Broad Walk; as I watched one of them toiled up bearing in triumph a crumpled paper bag. After incorporating it in the nest he flew down again to a wastepaper basket by the Palace and dragged out some more paper; the ground at this point was littered with paper apparently pulled out of the basket by the prospecting jackdaws. It abandoned the second bag, joined its mate on the grass and both soon flew up to the nest with indistinguishable small objects.[93]

---

93. ED: In *Birds and Men* (1951) EMN wrote: 'I have personally only ever seen three jackdaws in the heart of London, one of which, probably an escaped cage-bird, sat on the roof of No. 10 Downing Street uttering low hoarse chuckles while a Cabinet meeting was being held underneath, during the economic crisis of October 1949.'

I have attempted in this book to avoid raising matters of controversy, which in a plain account of bird-watching are out of place. But in the case of the **carrion crow** it becomes necessary for this aim to be relinquished. It would be cowardly to use such an excuse for turning a blind eye to the greatest blunder in the protection of London birds, especially since the attitude I have to uphold happens at the moment to be highly unpopular, and since Hudson, who used to plead for it, is dead.

Nowhere in the British Isles except in London has the carrion crow the remotest chance of becoming a common and approachable species. Over large areas he is quite wiped out; in most parts he survives, but only in a remnant too hardened to such perils for gun or poison or trap to have any power against him. No other species is so universally and vindictively persecuted. His outlawry is not, like that of the buzzard, or the merlin, or the barn owl, flagrantly unjust. We outlaw him falsely on high moral grounds as a villain (for he is no more so than species like the stork or the secretary bird which we choose to call good because they kill the creatures we want to see killed) but he might quite honestly be outlawed because his activities must inevitably conflict with our own and because, worst of all, he is armed with enough intelligence to pursue them in spite of us. Where we wish to preserve chickens or game or any small helpless domesticated creatures it is difficult to tolerate the carrion crow, not because he is vicious, as we prefer to think, but because he happens to like to eat such creatures at an earlier stage than we like to eat them ourselves. But in the special case of the metropolis these arguments can hardly apply. In London proper neither game-preserving nor poultry-farming can be called widespread occupations: the conceivable loss to man which could be inflicted by any number of carrion crows is infinitesimal. And yet they continue to be persecuted for three reasons: first, because their persecution is now an ancient custom and is continued out of habit; second, because there are ducklings to look after in the Royal Parks; and third, because it is believed that to let them alone would be fatal to the other birds. The first reason is not worth refuting, but it is not a frivolous one: if you landed a few ignorant settlers in any uninhabited country they would at once begin to shoot crows, or any birds which looked like crows, out of sheer force of habit. The second is more serious. There is a natural apprehension that if the persecution of crows were given up the ducklings would suffer by it. I expect they would, but not to anything like the extent that appears to be imagined. There is almost always a crow on the island in the Serpentine, and the cover there is now extremely poor, but for two years running the tufted ducks have succeeded in rearing a splendid brood in spite of it—few of the semi-domesticated mallards fare so well. It must not be forgotten that, short of turning the place into a poultry-farm, there is a definite limit to the number of mallards which can safely live on the Park waters, and this limit tends to be hopelessly outrun by a rate of natural increase adapted to the hard and perilous existence of birds in a truly wild state. If we take it that they hatch on average at least half a dozen eggs to each pair, and make full allowance for unmated and sterile birds, at least four of those half-dozen have in some way to be eliminated by next season if the population, already unnaturally dense, is not further to increase. We see the crows take toll of the ducklings and we jump to the conclusion that if the crows were killed that many ducklings would be saved, but it is not so: they would simply be eliminated by an alternative method, which would not harrow our susceptibilities because we should be powerless to massacre frosts and bad weather and sickness and the other forms it would assume. We shall meet this fallacy again.

The ducks are a characteristic and admirable feature of the London parks, and it would be unreasonable to suggest that their interests should not be consulted. But when it comes to the pitch of needlessly slaughtering other birds, which to the true naturalist are also characteristic and admirable inhabitants, in the misconceived interests of the ducks, it may be well to point out that a policy of treating the Royal Parks as a glorified duck-farm belongs to the period of Forest Laws and the Tower menagerie, not to the age of bird sanctuaries.

The third reason for persecution, to which the Bird Sanctuaries Committee appears to subscribe, is based on several false assumptions. It is assumed, as in the case of the ducks, that the victims of crows are additional to the losses from 'natural' causes: in reality the majority of them would perish just the same whether the crows became their residuary legatees or not. They have to die anyhow: you do not abolish death simply by abolishing one of its minor causes which happens to strike your purblind eye—you only alter the manner of it. To assume that as many as the crows take would otherwise live is quite unjustifiable, unless the species concerned is actually declining below its proper state. It is further assumed that if crows are not persecuted they will increase till they became a plague. There is no justification for that belief. The checks on the increase of birds of prey may be less obvious than the checks upon their victims, but they are not less powerful. The same argument is used to justify the slaughter of hawks, owls, cormorants and others as well as crows; those who use it forget that all through the Middle Ages the utmost rigour of the law was invoked to increase the supply of falcons for sport, and to molest them was to expose oneself to drastic penalties summarily enforced, but they never multiplied. The jealousy of such birds for their breeding-places is in itself, as a rule, quite sufficient to ensure that the number of pairs on a given area will never seriously increase, for interlopers and the young of previous years will be driven out by the watchful territory-owners far more efficiently than by the gun. But above all this misconception comes from a deficient sense of proportion. Outright slaughter by birds of prey is among the rarer (and also the most desirable) forms in which a bird is likely to meet death, either in Hyde Park or anywhere else. Because we see it happen openly, while the working of the others is hidden, we are persuaded to exaggerate its frequency to an altogether ridiculous extent. The fierce spectacular predatory creature snatches for himself no more than stray fragments of the colossal stream of quickly-consumed life which is perpetually being emptied into oblivion.

This mistaken policy has sprung simply from ignorance of the processes of nature. Lately, however, there has been something more—the ethical point of view. Over there, beyond Park Lane, when we wish to relieve people of their belongings we no longer present a pistol to their heads or board their treasure-ships with cutlasses. We do not even see whom we rob: it is a highly civilised community and all is done with pen and paper by commercial and financial transactions. This way of going about it has one relevant consequence: the old merchant-adventurer, or pirate as we should call him, could not, unless he was Robin Hood, lay claim to any moral intentions, but his successor can and does. In fact he has to, for a politician or a businessman or a financier who confessed to having as little moral principle as he has moral practice when it comes to the point would be denounced by a scandalised community. But the crows in Kensington Gardens have not yet reached this ethical plane. When they kill and eat a sparrow it never occurs to them to make a show of doing it entirely for the sparrow's good. They have not even improved on the crude old practice of consuming one sparrow at a time. If only they could devise a method of feeding not on one complete sparrow but on a portion of each of two hundred sparrows a day, till the whole number was consumed, they would put themselves in harmony with our enlightened sentiments and all would be well. We should no longer feel morally impelled to execute them.

But these are negative arguments, to show the falseness of the reasoning by which the slaughter of the London crows is justified. It must not be thought that there are no positive ones. The crow is the grandest, the most intelligent and the most devoted to his mate and young of all the wild birds that survive in the metropolis. There is hardly another species whose vanishing would leave such an irreparable gap.

In Kensington Gardens at the beginning of July last year (1925) I watched a family of crows which probably belonged to the nest near the Round Pond. The three young birds, rather jackdaw-like in build, were almost as long as their parents but much more slender and less hammer-headed, and probably only half as heavy. They seemed to fly better than the adults, with many sudden turns, showing considerable skill in manoeuvring, and a much more rapid beat—the flight, like the appearance, resembled a jackdaw's more than a crow's. Although they were able to find food for themselves (and I more than once saw two of the youngsters solemnly exchange presents of it, perhaps in imitation of the old birds) they still depended mainly on their parents, both of whom looked after them exceedingly well. If the young bird was at a distance when the find took place he would run trippingly over with an almost wader-like gait, in the end opening his wings and flapping them; reaching the parent he would halt, open his bill wide, screen with his wings like a hawk on its prey, often flapping them, and utter a low harsh sigh of pleasure, sometimes like a large husky starling, sometimes with a peculiar laugh not unlike a distant greater blackback, and twice with a quick 'Jack-jack-jack-jack-jack', scarcely distinguishable from the jackdaw's burst of chattering. The old birds, who kept complete silence, walked, or when in a hurry broke into a series of hops; in both shape and movements they must be the most ungainly of all well-known British birds. But their size is impressive and they carry themselves boldly erect on legs which are both stout and long; a woodpigeon looks a small bird beside them and stands less than half as high. Here, watching the young and old together, I was struck by the much better and almost graceful carriage of the young: it seems that they grow more ungainly as they mature. The field was parched after an abnormal drought; their employment was undoubtedly harmless and probably beneficial. I saw one of the adults pull up and give to the young a fat short corpse-yellowish grub. This family remained constantly together until the winter. One of these Kensington Gardens crows had a conspicuous patch of ashy grey on the secondaries and gave the impression of being a descendant of one of those mixed matings between the hooded and carrion crows which have frequently been recorded. On Christmas morning three of them were unusually loquacious and I saw one settle on a pinnacle of Kensington Palace and utter three deliberate caws—no more. On such an occasion that must have been an omen of the deepest die, but its significance was lost on me, and is still.

One morning in June I watched a crow strutting about the grass near the Round Pond. He came close to a pair of mallard but retired again, then proceeded aimlessly along in a sort of zig-zag, till about the fourth tack he sprang suddenly without warning about a foot aside from his course, landing beneath a chair with a sparrow firmly gripped in his claws. These and the powerful bill made short work of it, and the crow, coming into the open, lacerated his prey and bore it to the lower branch of a tree, retreating higher as I advanced and in the end flying away, always carrying the sparrow conspicuously in his bill. At the first cries a cock-sparrow, perhaps the victim's mate, flew down from a tree and, alighting on the chair, sat watching. Whatever his connection with the unfortunate he accepted the deed in a spirit which might be recommended to birdlovers who let their feelings run away with their understanding. He did not attempt to make a tragedy of it. To suggest what passed through his mind would involve too great a risk of anthropocentrism, but we of all generations ought to recognise that the impression death makes on the living is greatest when it is most rare, and ceases to produce its effect on those who are continually mixed up with it. Where life is normally brief, and above half a dozen years must count as old age, we have no reason to believe that such accidents are long remembered, or that the mind of an average bird is capable of giving it any lasting distress of that kind. With certain solitary species like the robin, and particularly with the larger more intelligent kinds, which pair for life and may live together for many years, the case is very different: no one who has tried to understand the mind of a bird is likely to deny that we inflict infinitely more agony by depriving a crow of his mate than the crow could inflict by doing likewise to any number of smaller and mentally less developed species. No birds are more devoted or more constantly in company than a pair of crows.

There is a further, and an entirely different, reason why the crow should be given an armistice in the London parks. W. H. Hudson, who understood the spirit and the mechanism of nature better than any other man has done, pleaded passionately for them, yet in the same week that a memorial to him[94] was unveiled by the Prime Minister in Hyde Park I saw an unsportsmanlike, and fortunately unsuccessful, attempt to shoot on the nest the only pair of carrion crows in Kensington Gardens. This nest was in the very summit of a dense and lofty horse-chestnut close to the Round Pond. It was about seven o'clock in the morning when the attempt took place and the keeper, having missed with both barrels, glanced at the birds mounting in alarm and, recognising that there was no hope of a second chance, walked briskly off past *Physical Energy* and disappeared.

In the notorious controversy which began to rage immediately after the unveiling, the question whether Hudson himself would have appreciated Epstein's *Rima* was hotly debated.[95] That may have been open to doubt, but no person with any knowledge of his character is likely to deny that he would have much preferred the sparing of park crows to any stone monument. That a persecution over which he expressed himself so strongly should be continued

---

94.   ED: The Hudson Memorial is a bas-relief monument to the writer and field naturalist, who had died in 1922. It is situated north of the Serpentine and was carved by Sir Jacob Epstein in the form of a figure of Rima—a symbolic figure of a bird girl in Hudson's romance *Green Mansions*—surrounded by a formalised flight of birds.

95.   ED: See, for example, E. Dancy 1937, *Hyde Park*, pages 136–137. The unveiling of the monument provoked bouts of philistinism and the bas-relief was defaced several times by green paint.

without comment in the very place of his memorial comes perilously near hypocrisy.

In early spring the courtship of the **woodpigeon** is a conspicuous sight in the Park and Gardens. The gliding love-flight is performed with slowly-beating wings and the bird every now and again ceases to move them, at the same time depressing his tail so as to rise a little and stall. The stalling is often accompanied by a triple clatter like the instrumental alarm heard when the bird is startled from a tree, which is not caused, as it is often imagined to be, by the wings colliding with the branches in its haste. This part of the ceremonies is practised chiefly in the open, for example above the Long Water. Display proper may be watched during April and May with an intimacy which can hardly be possible outside the middle of London. The cock behaves very like an overgrown domestic pigeon, bowing so low that his neck (puffed out till it becomes as short and stout as a pouter's) rests almost on the branch or the ground where he stands, with his raised tail fanned astonishingly wide, and cooing voluptuously all the time. Up to the end of August the gliding nuptial flight may still occasionally be noticed. By February fighting is in full swing. The combatants make great efforts to clash in the air with their feet free and in a suitable position for striking. The clashes are loud and spectacular, but I have never seen any serious damage come from them.

In Kensington Gardens I have seen a woodpigeon sitting by the first week of March, but the middle of May seems to be the period when the majority begin nest-building. While the hen is incubating the cock sometimes bears her company, sitting on the nearest branch. The method of feeding the young is strange, and disgusting. A young bird of almost mature appearance, except that it lacks the white 'ring' mark on the neck, is slenderer, especially about the head, and has a duller breast, often turns out to be still unweaned. One of this stage which I observed near Speke's Monument followed its parent with constant importunities till it was allowed to settle on the branch alongside, screening its wings like a hawk and thrusting its slender head inside the old bird's open bill so far that its eyes went almost down the throat. Then both would work up and down as regularly as the handle of a pump, with tremendous heavings and tuggings and wrenchings. It was clearly an exhausting process for the parent, who became at intervals breathless and half-choked, and kept escaping to another tree, to which the insatiable fledgling always followed till its importunity persuaded the old bird to consent to a further application of the stomach-pump.

Greed is a principal characteristic of the London woodpigeon. It is that alone which has made him relinquish his native unprofitable shyness. He cannot bear to see other birds sharing the feast, however much there is of it, and though he may be too occupied to drive them away he never ceases to make a grievance of their intrusion. He will abandon his own crust to bully a sparrow into yielding him another and I have seen two woodpigeons fight over a crust when the ground on all sides was strewn with them. He enjoys the most comfortable existence of all the London birds. Unmolested by man and too large to stand in fear of other enemies he finds the bread and grain which are so lavishly distributed closer to his normal diet than any other species; his burly form and brazen manner secure him (except at the water's edge) the first option on supplies, and trees are too plentiful for any change in his usual habits to be forced upon him. He is not, as the ordinary pigeons are, annoyed by the intrusion of hordes of intolerable starlings: he lives the untroubled existence of a domestic dove without the slightest loss of freedom. That he thrives on it is clear from his greatly increased bulk: in winter, when country woodpigeons penetrate to Hyde Park, they are picked out not only by their shyness but by seeming little more than half his size. Far more than the sparrow, which has much to put up with, he finds his earthly paradise in London.

Moorhens flourish on the Serpentine, the Long Water, the Fountains, in the Dell, and in the Dutch Garden by Kensington Palace, but not on the Round Pond. The parks do not contain a species more interesting to watch. There is a dramatic quality about all their movements; even their gait is tense and arresting. In spring, when love and hate consume them, battles are desperate and prolonged. For days the struggle rages, with hardly an interruption while the light lasts—they snatch their meals in momentary lulls. In one such conflict, which took place on the lawns by the Long Water, close to *Peter Pan*, a small but determined male was in possession of a mate and territory, while the attacker, fully half as heavy again, threatened to wrest both from him. To make things worse his mate was in league with the invader. While the fighting lasted she openly encouraged the enemy, beating a sort of tom-tom accompaniment with her low mechanical 'Eck-eck-eck-eck', repeated at brief intervals; and when at a breathing-space in the battle her mate marched proudly up to her she kicked him savagely, evidently furious at his success. Then, twisting her neck without turning round, she gave him (it seemed to me) such a look of intense hatred as one is rarely able to observe in birds. In the swift running fights with taut outstretched necks (of which most of the skirmishing consisted) the lover, despite his superior size, was almost invariably the fugitive, and it was his tail which suffered most of the stabbing pecks, probably less painful than ignominious. He fled with long graceful strides, helping himself at times with flickering, rhythmic beats of his broad wings, which the swifter pursuer never found it necessary to use. Sometimes the insistent tom-tom of the hen taunted or encouraged him to the point of turning to bay, leaping into the air and aiming futile blows with his pale green feet like a half-hearted fighting-cock. To show how little dread he had of his pursuer he would stand and crow defiantly in his very face—a bravado of which he lessened the effect by turning tail and taking to his heels (or even, when hard pressed, to his wings) at the next onslaught. Time after time he was routed and driven away, but always he would attempt to dodge the defence and win his way to the hen's side. The more excited her half-whispered encouragement made him the more determinedly her mate repulsed his advances. It was strange to find the heavier, handsomer bird, who was into the bargain the attacker, consistently avoiding a pitched battle, in which his superior weight ought have given him the victory, but the emotions of breeding birds form such potent stimulants that whatever the odds an immense psychological advantage always rests with the defender of the home.

These lawns near *Peter Pan* are amongst the best places in London for watching moorhens at close quarters. There are often more than a dozen present, and they are quite unafraid. March is the time of fiercest rivalry, and it was in the middle of a morning early in the month when I saw the strangest battle that I have ever watched. Out on the Long Water, about five yards from the bank, were three moorhens. Two were in a most fantastic attitude, face defiantly to face, sitting bolt upright in the water out of which their breasts and heads emerged in the posture of a seal or a mermaid.

Their wings, fully spread, rested almost submerged and half-reversed, with the front of the wing held above and in rear of the actual rear margin. Their long skinny-faced green toes were planted firmly on each other's breasts, and there they sat, tense and motionless. Occasionally a violent wrestling match took place, the object of which seemed to be to kick the feet up and bring them down on the adversary's head so as to press it under water, which one of them during a fierce bout actually succeeded in doing. The third bird kept up an incessant tom-tom accompaniment right in their ears, and drove off a fourth which at one stage tried to approach. The end of the struggle came suddenly after a long spell of dogged wrestling: in one of the great onslaughts the bird who had been losing on points was seized with a panic and turned tail—he was not given the opportunity to make a fresh stand, and dived after a moment's pause, the victor retiring with the noise-maker, who seemed to be his mate. Though there was a clear temptation to give chase he was in the end allowed to retire in peace; I never saw a bird look more thoroughly soused.

If the behaviour of the old birds is best seen on the lawns by the Long Water, it is on the smaller pools at the Fountains, the Dell in Hyde Park, and the Dutch Garden between Kensington Palace and the Orangery, that their domestic affairs allow the most intimate observation. On the first, when a brood is sheltered among the tub-grown rushes, the parents become carriers rather than foragers to the young: they ferry ceaselessly to and fro accepting food from the Londoners at one end and dropping it into hungry mouths at the other. One May afternoon, hearing a shrill querulous cry in one of these basins at the head of the Long Water, I saw a young moorhen all alone and very conscious of it. Swimming round he came in the end to the sloping plank and walked up it, looking like a jacana with his grotesquely vast *Struwwelpeter* toes, and long slender stilt-like legs. His plump downy body was black and bore no more reasonable proportion to the legs than a fresh-dropped colt's; the violets and carrion-reds of the crown gave at close quarters a curious and rather repulsive impression, like the vivid patches on certain monkeys. Stumbling awkwardly along the stone edge he was alarmed by some children and leapt readily into the water like a frog, though the drop of eighteen inches was quite a dozen times his own height. The parent, who had now heard his cries, hurried over from the other basin.

The moorhens at the Dell gained a sudden fame in the May of 1925, when some herons, pinioned birds brought from Richmond Park, seized their nest, enlarged it with sticks and began to incubate the eggs. Apparently they lost interest in it soon, for a day or two later the moorhens had resumed undisturbed possession. It was amusing to see one of them, after sitting patiently in full view of an interested public, slip nimbly away on hearing the footsteps of a gardener who came down to fill a watering can and look at her nest. She hid in the reeds with only her red bill showing, but slipped back the instant he turned away; the little comedy, invisible to him, was plain to those behind the scenes on the opposite bank.

But moorhen-watching is one of the rare aspects of bird-watching in London which is already widely appreciated, and which not many birdlovers can be accused of having neglected. To write of them further would be wearisome to those who very likely know the bird better than I do, but these slight incidents from the Dutch Garden may serve to bring this long chapter to an end. The first was a moorhen standing under one of the tanks so that the slight trickle of overflowing water played refreshingly on its back; another bird flew up from the other side and, launching itself across the tank, impelled a great wave over the outfall end, so that the voluptuary found his shower-bath turned suddenly into a Niagara. This was probably unintentional, but a little later, when a moorhen scrambled up into a clump of rushes and made its way gingerly out above the water as far as their stems would bear its weight, another followed to drive it off; naturally the retirement of the first bird still farther out and the added weight of the second caused the rushes to bend so much that the first was tilted off into the water. Another in a similar position was cleverly brought down by a swimming bird, which came up and gave the tips of the rushes such a sharp tug that the percher lost his balance.

Chapter V

A Bird Census of Kensington Gardens

In America the idea of taking a bird census has experimentally been put into practice.[96] In England, where the greater variety of the landscape in small areas would give more interesting results, little or no substantial work of this kind has so far been undertaken. Yet the researches and observations of Mr Eliot Howard[97] and others have lately made it clear that territory is a more vital factor in birdlife than has ever before been suspected. The insectivorous birds, at any rate, have been shown to occupy during the breeding season a considerable patch of ground for each pair from which all others of the same species are jealously excluded. Without such a system it would often happen that twenty broods of insatiable nestlings would hatch out simultaneously on an area incapable of supporting more than ten, with the result that all would be undernourished or perhaps starved. There would be nothing to secure the separation of nests by a sufficient distance. The value of a bird census in providing data for further investigation of the part played by territory needs no further emphasis.

Eliot Howard and Chance in Worcestershire, and Burkitt in Ireland, have undertaken studies of small areas with a definite object in view—Chance's was incidental to a study of cuckoos[98] and Burkitt has confined himself to the robin[99]—while Alexander at Tunbridge Wells carried out a plan of mapping nesting areas but limited the scope to summer migrants.[100] Thus, while the subject has several times been touched on incidentally, it seems never, in this country, to have been tackled without prejudice: the taking of a bird census upon an adequate scale is work that still remains undone.

---

96. ED: See 'Census of Songbirds' on page 86 of *The Audubon Society Encyclopedia of North American Birds* (J. K. Terres 1980).

97. ED: E. Howard 1920, *Territory in Bird Life*.

98. ED: E. Chance 1922, *The Cuckoo's Secret*.

99. ED: See footnote on page 31.

100. ED: See Chapter 4 'The Kentish Weald: Population and Territory' in *Seventy Years of Birdwatching* (H. G. Alexander 1974).

A Bird Census of Kensington Gardens

During the winter of 1925–26 I took, with the help of my brother, a complete census of the birds of Kensington Gardens. It is far from being a normal area, and if it can be termed representative at all it is representative of only a very small fraction of the land surface of Great Britain. But for census purposes it has some rare advantages. Except on the east, where Hyde Park touches it, it is practically an island in a sea of houses. Its bird-life is distinct and isolated, not confused by passers-by from an adjoining countryside. It is also a place where feeding of the birds by man is carried out on an enormous scale—so regularly and in such quantities that a great part of the bird population is dependent on it, and is consequently very approachable.

The general idea was to repeat the count at intervals of about a month, with a view to ascertaining the losses sustained during the winter months and how far these were caused directly by severe weather. Owing to unexpected difficulties, and the fact that no appreciable change seemed to have taken place, the February count was abandoned, but at the opening of November, December, January and March censuses were successfully taken.

The first census was made on 2nd and 4th November 1925. It might have been possible to complete it in one day, but the working time (even with two observers) was nine hours without allowing anything for rests and the amount of walking involved altogether incredible. That was, therefore, scarcely practicable and certainly not desirable, seeing that it would have meant prolonging the count into the hours when movements to or from the roost are in progress.[101]

It will be as well at this point to deal with the whole subject of accuracy. The obvious question is: how do you know that the same birds were not counted two or more times, or that some by a simple flight did not escape the census altogether? The answer to that must

---

101. ED: EMN described how he and his brother conducted the census in an article entitled 'How we counted the Cockney sparrows' in the 1st February 1928 issue of *The Evening News*: 'We divided the Gardens into 19 sections, according to the lay-out of the paths and other obvious boundaries; wherever possible these took the form of long narrow strips which could be covered in one sweep by the two of us with field-glasses working abreast. The numbers of each species met with were noted down at the time, and on finishing each section were added up to be entered into the proper division of the sketch map, which had a stiff card backing. As the trees were bare next to nothing could escape a close scrutiny, and in most cases the birds were not shy enough to be driven on ahead by our approach.'

be that a bird census cannot claim the same degree of accuracy as a human census in a highly-organised country like Great Britain. All that can be done is to observe very carefully beforehand what considerable movements take place in the course of a day—an example from the Gardens is the concentration of sparrows in the late morning and afternoon around the Fountains and *Peter Pan*—and take them into consideration, so that if the area which loses birds is counted before the movement, that which gains is counted also before any movement takes place and not after, when many would be included twice. These are precautions of common sense, but after all have been taken there is still some coming and going of birds across the arbitrary boundaries. It has therefore to be assumed that movements are of two kinds: general ones taking place at regular times, which are known and provided against, and fortuitous comings and goings (not affecting a very high proportion of the bird population) which roughly cancel one another out. Here is one possible source of error, for if by any chance there is an unexpected general movement—an unlikely contingency fortunately, if the preliminary observation has been thorough—it is impossible to cope with. Another, obviously, lurks in the mere process of counting. It would be extravagant to claim that not a starling or a woodpigeon in the treetops, let alone a sparrow in the shrubberies, has been overlooked. But practice attains in a method which appears difficult and unlikely to be trustworthy an unexpected degree of accuracy. Another example of this is the well-known skill of the Australian sheep-counter. The number of birds supposed to be occasional visitors to the Gardens—kestrel, redwing, coal tit, pied woodpecker and others—unearthed in the course of the census was an indication of the efficiency of the search.

I do not, therefore, claim that the census figures as a whole are without a possible error of up to twenty per cent, although I personally am convinced that they represent within ten per cent, and in the case of the two complete counts within five per cent, of the exact population as it was at the times stated. The more nearly instantaneously a census could be carried out the less danger of error could arise, but even to complete such a small area as Kensington Gardens within an hour would require at least sixteen thoroughly competent observers. Fluctuations may be very rapid—three consecutive counts at the Round Pond gave the number of gulls present as 170, 155 and 215 within a very few minutes—and even in an hour there would be at least a theoretical possibility of confusion; the advantage of such a number of observers would lie less in the speed of counting than in the fact that it could be repeated three or four times in one day. If the results then showed any serious discrepancies the fact would be more likely to lie in the instability of the bird population than in the census-takers. The greater dangers of error in a count spread over two days are more apparent than real: it is arguable that by this method one obtains a more reliable result than by any attempt to secure an instantaneous view of the bird population. Any census is bound to give an illusion of exactness to what is in reality ceaselessly fluctuating: statistics may make it appear that the population of Liverpool stands for ten years at 746,421 and then jumps in a night to 802,940, but in practice obviously two exact counts on the same day might show a discrepancy of up to several thousand.

The November and December censuses were both complete, and the greater part of the period between them was abnormally cold; when the second was taken the Round Pond was frozen over, and had been for many days. The results of these two counts were as given in the table overleaf. The special points raised by the table may be briefly dealt with. The extraordinarily slight fluctuation in the number of sparrows proves how little such birds are affected by cold so long as their food supply continues. We noticed only one dead sparrow, leaving, from the unpractical standpoint of pure mathematics, seven others unaccounted for. But the closeness of these totals masks a very considerable upheaval. In November there had been 595 in the sections west of the Broad Walk; in December there were only 238. Across the Broad Walk there had been an equally striking redistribution: the 'back blocks' of trees were everywhere almost emptied, and the neighbourhood of the Round Pond and Long Water had attracted large numbers; the increase on the banks of the Long Water was from 499 to 825.

A similar but less remarkable redistribution of starlings was brought to light. In the border sections (totals 135 in November, 143 in December) they were very steady, but in the central and north central divisions they fell sharply from 219 to 76 and in the southern group they rose remarkably from 57 to 124. The great majority were feeding on the grass in sunny open parts; they are

BIRD CENSUS OF KENSINGTON GARDENS

Species	First Census 2 & 4/11/25	Second Census 3 & 5/12/25	Change
House-sparrow	2603	2595	-8
Starling	411	343	-68
Black-headed gull	289	440	151
Woodpigeon	241	226	-15
Mallard	240	9	-231
Moorhen	26	20	-6
Blue tit	37	17	-20
Great tit	19	18	-1
Blackbird	21	15	-6
Robin	16	12	-4
Pochard	16	0	-16
Tufted duck	7	0	-7
Skylark	11	0	-11
Redwing	0	15	15
Common gull	0	14	14
Chaffinch	10	9	-1
Throstle	4	10	6
Dunnock	6	7	1
Wren	6	4	-2
Jackdaw	2	8	6
Mistle-thrush	3	1	-2
Greenfinch	2	1	-1
Carrion crow	5	4	-1
Stock dove	0	7	7
Coal tit	3	0	-3
Pied wagtail	1	2	1
Gadwall	1	0	-1
Kestrel	1	0	-1
Pied woodpecker	1	1	0
Brambling	0	1	1
TOTAL	3982	3779	-203

little dependent on man. The redistribution may reasonably be traced to the fact that, whereas the central and northern areas are level, the southern have a sharp fall towards Kensington Gore, and so catch enough sun to be easily thawed when the other part is frozen. The greater softness of the ground on this slope was very noticeable. Taking the figures at their face value about 145 starlings were driven out of the central and northern divisions, of which 67 found room in the southern, 8 on the borders and the remainder were expelled from the Gardens altogether, perhaps no farther than Hyde Park. This is an interesting instance of the effect of frost on the starling population, and it shows how in hard winters, like that of 1916–17, great numbers of the normally stationary native race are frozen out and migrate (not south but westwards) looking for softer feeding-grounds. For example, a starling marked in Berkshire in 1913 was recaptured at home in February 1915 but finally in Cornwall in the arctic February of 1917.

The prevailing weather, an unbroken frost of more than a week's duration, accounted for the wholesale desertion of mallards, pochards, tufted ducks and the gadwall, which fell by 255 all told. The increase of 165 shown by the more aerial and mobile gulls is significant and cannot be treated separately: the contrast may be read to mean that fine cold weather may freeze out the ducks without diminishing the supply of offerings from human visitors, and the removal of their competition leaves the gulls with a monopoly and results in an increase of their numbers. This after all is only statistical evidence of the intensity of the struggle for existence between the two forms which can be observed in progress at any hour of the day. The number of gulls on the Round Pond, varying from hour to hour if not from minute to minute, is not, as it seems, fortuitous: it is adjusted by the interplay of complex forces of supply and demand as exactly as the price of butter.

The larger birds do not (except by migration of the ducks to the Serpentine) show any traces of having suffered badly through the hard weather. Nor do the sparrows, which are dependent on man. But with the smaller insectivorous kinds the case is very different. The decline of the starling has already been described. The three tits lost 24 in the month out of a total strength of only 59. The blue tit was the worst sufferer, with a drop of more than fifty per cent; the larger, sturdier great tit was scarcely affected. The London parks are apparently among the few places where no one makes any attempt to feed these birds. The blackbird also lost thirty per cent and the robin twenty-five per cent. On the other hand, the appearance of 15 redwings where none had been found before, and of seven stock doves and six new throstles, pointed to a slight influx of refugees from other parts. The jackdaw is a resident but several of the birds wander freely in the dead season and the full strength—now four pairs—was not invariably present.

On 3rd January 1926 the third census fell due. It was not planned on the same scale as its predecessor, for the data already obtained were enough to indicate which sections contained the highest proportion of the total population and which were most sensitive to change, and the labour involved, difficult for any observer who has not attempted such a task to appreciate, discouraged unremunerative effort. The area selected stretched right across the Gardens from the Long Water to the south-west corner. Before the count began it started to drizzle, and the rain getting into its stride poured down without intermission for the rest of the morning. All that could be said was that for part of the time it was not nearly so hard as it was for the other part. This did not tend to make the business pleasanter, but the changed conditions were at least interesting. For ten days now the weather had been consistently rainy and the parks comparatively little frequented. In the 12-hour periods from Christmas Day to New Year's Day inclusive the rainfall at Kensington Palace had been 1.5 mm, 3.8, nil, 1.4, 3.8, 1.5, 0.9, 5.8, 1.0, 7.5, 1.4, 3.0, 0.9—total 32.5 millimetres or about 1.3 inches, very evenly spread. As I expected, the numbers of sparrows in the crowded sections, where in better weather food is lavishly provided, proved to have fallen catastrophically; in other sections the fall was only slight. Blackbirds, robins, starlings and tits, which had suffered most before, were flourishing; ducks were more scattered but not seriously reduced. In order to save space the table for this January census is combined with the next and final one, taken two months later, on 3rd and 5th March. (The figures for the 3rd affect only the two check-counts for the Round Pond.)

The 5th of March came at the end of almost a week of generally fine weather interrupted occasionally by sudden storms and even blizzards. Although their retirement from London was now due, gulls, pochards and tufted ducks maintained or even exceeded their normal numbers. This, at any rate in the case of the ducks, pointed to an influx of birds on passage.

COMPARATIVE FIGURES FOR CERTAIN SECTIONS—LAND BIRDS

Species	XV & XVI (By Long Water) Nov	Dec	Jan	Mar	X (Behind *Peter Pan*) Nov	Dec	Jan	Mar	XI & XIX (Back Blocks) Nov	Dec	Jan	Mar	Aggregates Nov	Dec	Jan	Mar
Sparrow	447	636	225	503	249	248	236	368	292	258	133	109	988	1192	594	980
Starling	9	0	6	3	53	21	78	74	63	14	79	4	125	35	163	81
Woodpigeon	7	17	3	31	43	20	59	36	22	4	12	22	72	41	74	89
Blackbird	5	3	0	4	0	1	4	2	2	0	4	0	7	4	8	6
Robin	4	2	2	0	0	0	0	0	0	0	0	0	4	2	2	0
Great tit	5	0	0	1	2	3	0	3	5	3	4	0	12	6	4	4
Blue tit	2	1	0	2	3	3	3	5	6	1	5	0	11	5	8	7
Chaffinch	0	0	0	1	0	3	1	6	9	0	1	0	9	3	2	7
Throstle	–	–	–	–	–	–	–	–	–	–	–	–	2	1	4	1
Dunnock	–	–	–	–	–	–	–	–	–	–	–	–	2	0	0	2
Pied wagtail	–	–	–	–	–	–	–	–	–	–	–	–	0	1	1	0

MARCH ONLY: redwing 6; carrion crow 4; mistle-thrush 2; kestrel 1

Comparative Figures for Certain Sections—Water Birds

Species	Long Water				Round Pond				Aggregates			
	Nov	Dec	Jan	Mar	Nov	Dec	Jan	Mar	Nov	Dec	Jan	Mar
Black-headed gull	53	147	12	51	215	287	255	195	268	434	267	246
Common gull	0	3	1	2	0	11	14	9	0	14	15	11
Mallard	35	4	49	65	205	0	146	109	240	4	195	174
Tufted duck	0	0	0	0	7	0	5	46	7	0	5	46
Pochard	0	0	0	0	16	0	18	41	16	0	18	41
Moorhen	24	19	30	18	0	0	0	0	24	19	30	18
Coot	0	0	4	0	0	0	0	0	0	0	4	0

Total Sparrow Population

Nov	Dec	Jan	Mar
2603	2595	1840	2505

(The figures for January and March are estimates based on incomplete counts; other methods of estimating gave 1775–1813 for January and 2690 for March.)

The results of the census led me to the following conclusions. Kensington Gardens, where bread is given away in quantities instead of having to be searched for, are such an outstandingly attractive habitat for the particular species which flourish in them that whatever gaps may be caused by mortality are immediately filled by an eager pressure from outside. The total population probably remains pretty constant at all seasons in normal weather; when an exceptionally rainy spell occurs and people cease to find much to attract them in the sodden park, the food supply drops suddenly and an exodus takes place, probably in proportion to the length of the spell. That many wait to be starved out is unlikely; but the reputation of the Gardens as a land flowing with milk and honey, far from abolishing the struggle for existence, appears to intensify it, or rather to cause an abnormally bitter struggle for Easy Existence (which may humanly be compared to the effects of a 'gold rush' in America).

The bird populations recorded cannot be taken to throw any light whatever on the normal density of bird population in Britain,[102] owing to the abnormality of the circumstances. The area of the Gardens is given as 275 acres,[103] and the average density in November was therefore 14.5 (over 7 pairs) to the acre. In December it had fallen to 13.7. But the distribution was very uneven. On the Long Water in November there were 112 birds to 9 acres—about 12½ to the acre or well under the average for the whole Gardens—but on the Round Pond there were 444 birds to 7 acres or 63.4 to the acre. A month later, when the Pond was frozen over, there were only 298 (still over 42 to the acre); in January there were 438 (62.6 per acre) and in March 400 (over 57 to the acre). The highest density on the Long Water was 19 to the acre—a striking contrast, since this last is intrinsically the more favourable for bird-life, having very much better cover and feeding-grounds. The explanation, of course, is that the Round Pond is accessible to the public on all sides and much more suitable for feeding water birds; the general ambition is not to hunt but to be fed.

---

102. EMN: From subsequent investigation I have formed the impression that this in favourable circumstances stands at about six to the acre. [ED: See EMN's 1988 paper 'The bird population of Great Britain' in *British Birds* (81: 613-624).]

103. ED: 111 hectares; the area of Hyde Park is 350 acres (140 hectares).

A Bird Census of Kensington Gardens

The greatest density of bird population recorded anywhere by the U.S. Bird Census was on a suburban estate near Washington— 135 pairs of 40 species on 5 acres, or 54 to the acre, but these I believe were permanent inhabitants and, in any case, the number of species represented was eight times as great as on the Round Pond. (In British Guiana, Beebe[104] counted 464 *species* of birds in a quarter of a square mile of tropical jungle.) In the absence of statistics for comparison it is difficult to say whether the winter density of bird population on the Round Pond could conceivably be equalled without artificial encouragement. In any case the maximum density in Kensington Gardens was not here but in the long narrow division west of the Long Water where, on an area which cannot be more than five acres at the most, there were in December 659 birds and in March 545—the densities being in the region of 132 and 109 birds per acre respectively, if not more. Conversely, there were large areas very sparsely populated with birds.

My impression was that the bird population had become artificial and dependent on man to such an extent that, out of less than 4,000 birds, more than 3,000 would be starved or driven out if the Gardens were suddenly closed to the public. The five leading species numbered in November 3,784 (all but 198 of the whole population) and, of these, the house-sparrow, mallard and black-headed gull are almost entirely, and the woodpigeon and starling largely or very largely, dependent upon man. Without him I doubt if the Gardens would support more than about 60 of the 2,603 sparrows, 15 of the 289 gulls, 30 of the 240 mallards and perhaps as many as 90 of the 241 woodpigeons and 100 of the 411 starlings. That, of course, is nothing more than an opinion, based on the numbers of these species present on roughly similar estates where the birds are not fed: it takes no account of the fact that if the ground were used for other purposes (as housing estates or to grow crops, for example), or even if it were left entirely to its own devices, it would gradually make good part of the loss by attracting a fresh avifauna. As to the gulls, it is doubtful whether any would spend much time in Kensington Gardens if there were no people.

---

104. ED: William Beebe, then Director of the Department of Tropical Research at the New York Zoological Society, which established a tropical research station near Bartica in British Guiana. His accounts of the research station's work appeared in book form in *Tropical Wild Life in British Guiana* (1917), *Jungle Peace* (1918), *Edge of the Jungle* (1921) and *Jungle Days* (1925).

A Bird Census of Kensington Gardens

Professor Forel on Lake Geneva found the summer population of black-headed gulls to number about 300–500 and the winter (September–March) about 3,000, giving from one bird on 480 acres (summer minimum) to one bird on 48 acres (winter maximum), at which rate the Serpentine, Long Water and Round Water together would support precisely one gull. But some allowance must be made for areas of Lake Geneva too deep to be of much use to gulls.

Another interesting point was that, of the four most numerous species, three—the starling, woodpigeon and black-headed gull—retire to the last bird to roost in some other place during the winter months, the gulls to the Thames reservoirs, the starlings about Trafalgar Square, St Paul's and in between, and the woodpigeons to the Serpentine island or Battersea Park. The night population in winter must therefore be about a thousand less than the day population.

I should not end without acknowledging more fully the help given by my brother. He took part in all but the last of the four counts, and having to undertake that alone made me realise that without his skilful and ready assistance my project could never have been carried out. Although so far as its original object of throwing light on winter mortality was concerned the census proved a dead failure, the incidental results were, I think, of sufficient value to justify the labour of making it. But whatever may have been accomplished, the need for a bird census of some characteristic part of England, and if possible of more than one, remains unaltered. With the additional complications of arbitrary boundaries, fluid population and immensely greater area to cover, the difficulties may well discourage the most enthusiastic observer, but the occasion is there, waiting for ornithologists to rise to it.

Two Bird Censuses of Kensington Gardens, 1925 and 1975

By E. M. Nicholson and R. F. Sanderson

On 2nd and 4th November 1925 a complete census of Kensington Gardens was taken by EMN and the late B. D. Nicholson, being one of the first complete censuses to be made of any substantial area in Great Britain. Fifty years later, EMN, as a member of the Secretary of State's Committee on Bird Sanctuaries in the Royal Parks, enlisted the co-operation of a fellow member, Sir Con O'Neill GCMG, and RFS, of one of the Official Observers for this area, to repeat the census on the anniversary date, 2nd November 1975. In the unavoidable absence of the other Official Observer, D. A. Boyd, his place was taken by P. Ball. Having thus four participants against the original two, it was possible to complete the count within 5½ hours, between 8.05 a.m. and 1.30 p.m., totalling about 22 man hours, against some 18 man hours for the original count, which had to be spread over two days. Every effort was made to keep methods reasonably comparable, although it was not possible to reconstitute all the original subdivisions as the map demarcating them was missing.[105]

Since 1948 two intermediate censuses had been taken of Kensington Gardens, by S. Cramp and W. G. Teagle on 19th December 1948,[106] and a second by RFS on 19th November 1966.[107] Although neither of these was quite so early in the winter, they can be accepted as broadly comparable.

Recalling that even exact counts yield appreciably different results at different times of day, in different weather conditions, and at different seasons, the remarkable general consistency of the results over so long a period, and their close correspondence with what is known from careful everyday observation, give confidence that they can be relied on for a general picture of the bird population of the Gardens and its salient trends over the decades. Even species regarded as rare or intermittently present rarely fail to show up in the census and changes in bird population tend to be shared by more than one Inner London park at the same time. Kensington Gardens have undergone during the half-century some

---

105. ED: The 1975 census is mentioned in *Bird Life in the Royal Parks 1975*, which devoted page 5 to it, but otherwise the results have not been published, apart from the drop in counts of House Sparrows which was reported in *Lond. Bird Rep.* 40: 65.

106. ED: S. Cramp & W. G. Teagle 1950, A Repeat Bird Census of Kensington Gardens. *Lond. Bird Rep.* 14: 41–48; see also S. Cramp 1949, The Birds of Kensington Gardens and Regent's Park. *Lond. Bird Rep.* 13: 37–45 and S. Cramp and W. G. Teagle 1951, A Bird Census of St James's Park and the Green Park. *Lond. Bird Rep.* 15: 48–52.

107. ED: R. Sanderson 1968, The Changing Status of Birds in Kensington Gardens. *Lond. Bird Rep.* 32: 63–80.

A Bird Census of Kensington Gardens

significant but not very drastic changes. The layout and landscape have altered little, but the stock of tall mature and over-mature trees, especially elms, has been depleted by losses from disease, windblow and felling. The Broad Walk has been replanted with alien species, but elsewhere replanting has been inconspicuous. Several new structures have been erected, notably what is now the Serpentine Gallery and the elegant bandstand south of the Round Pond. Sheep have long since ceased to be brought in for summer grazing, and the sheepfold behind *Peter Pan* has vanished. Wartime allotments, bomb craters and most of the former nestboxes no longer exist. A new enclosed sanctuary has been created between Buck Hill and the Long Water Sanctuary, and very recently the Long Water has been officially closed to boating from the Serpentine. The effects of these and other environmental changes have not been inconsiderable, and have been added to by the changing habits and practices of those who use the Gardens.

The effects of these changes on the bird population are not clear but it seems probable that an increased area of infrequently disturbed sanctuary has helped especially the small insectivorous species while the loss of a majority of the holes and crevices formerly available in old trees has worsened conditions for woodpeckers, Nuthatches, Jackdaws and other arboreal species. Such indirect influences as reduction of air pollution and of horse traffic must also be borne in mind, as well as general trends such as the colonisation of Inner London by breeding Great Crested Grebes, Magpies, Jays and Long-tailed Tits. It seems unlikely in the light of present knowledge that specific habitat changes during the period have influenced the bird population as greatly as indirect factors.

With this introduction we first present in a comparative table the total figures for the counts of November 1925, December 1948, November 1966 and November 1975, indicating the number of species included on each occasion.[108]

	1925	1948	1966	1975
No. of species counted	26	24	28	43
No. of birds counted	3982	1949	2169	2319

---

108. ED: It should be borne in mind that these censuses were conducted on various days of the week: Monday 2nd & Wednesday 4th November 1925; Sunday 19th December 1948; Saturday 19th November 1966; and Sunday 2nd November 1975. On each of these occasions, except the last, full or partial censuses were also conducted on other dates: Thursday 3rd & Saturday 5th December 1925, Sunday 3rd January and Wednesday 3rd & Friday 5th March 1926; Sunday 23rd January and Sunday 13th March 1949; Saturday 17th December 1966 and Saturday 4th February 1967.

A Bird Census of Kensington Gardens

BIRD CENSUSES OF KENSINGTON GARDENS

Species	2 & 4/11/25	2/11/75	Change
Little Grebe	0	1	1
Great Crested Grebe	0	5	5
Grey Heron	0	1	1
Mute Swan	0	5	5
Canada Goose	0	59	59
Gadwall	1	0	-1
Mallard	240	246	6
Pochard	16	13	-3
Tufted Duck	7	142	135
Sparrowhawk	0	1	1
Kestrel	1	2	1
Moorhen	26	20	-6
Coot	0	29	29
Black-headed Gull	289	251	-38
Common Gull	0	16	16
Lesser Black-backed Gull	0	2	2
Herring Gull	0	1	1
Woodpigeon	241	386	145
Tawny Owl	0	1	1
Great Spotted Woodpecker	1	1	0
Skylark	11	2	-9
Pied Wagtail	1	11	10
Wren	6	13	7
Dunnock	6	17	11
Robin	16	14	-2
Blackbird	21	191	170
Song Thrush	4	7	3
Redwing	0	8	8
Mistle Thrush	3	4	1
Goldcrest	0	4	4
Long-tailed Tit	0	11	11
Coal Tit	3	3	0
Blue Tit	37	45	8
Great Tit	19	13	-6
Nuthatch	0	1	1
Treecreeper	0	2	2
Jay	0	8	8
Magpie	0	8	8
Jackdaw	2	0	-2
Carrion Crow	5	52	47
Starling	411	148	-263
House Sparrow	2603	544	-2059
Chaffinch	10	6	-4
Greenfinch	2	10	8
Redpoll	0	15	15
TOTAL BIRDS	3982	2319	-1663
TOTAL SPECIES	26	43	17

In 1925, four additional species—Common Gull, Stock Dove, Redwing and Brambling—occurred in the December count and both Canada Goose and Mute Swan were excluded as non-feral. In 1975 Bullfinch was seen next day. Disturbance was almost certainly much greater in 1975, but with more and better-practised observers finishing within one morning other sources of inaccuracy were reduced.

Feral Pigeons were not counted in 1975, but there were 619 in 1966; numbers in 1925 and 1975 would have been much lower, more nearly as in 1948 when there were 66. Ornamental geese, Mandarin Ducks and recently-released Rooks were excluded from the 1975 counts. As a mammalian predator the grey squirrel was formerly important, some 21 being noted in 1925, mainly between the Fountains and Speke's Monument; in 1930 a campaign to eliminate them began and in 1948 none was noted, but by 1975 there were at least ten.

Considering the gross results, the number of species recorded in 1966 was identical (at 26[109]) with that 41 years earlier, allowing for the previous exclusion of the Canada Goose and Mute Swan as introduced. The 1948 level was two fewer, the Wren and Greenfinch being missing regulars on that occasion.[110] In contrast 1975 showed a striking increase to 43, added diversification ranging from two grebes and a heron at one end of the list to Long-tailed Tit, Goldcrest and Redpoll at the other. Species represented by more than three individuals nearly doubled, from 18 to 31. Since a similar picture is shown in the regular records of the Official Observers[111] there seems no reason to doubt its validity, or to question the implication from the counts that it is not just a question of intensified observation, but represents a real change.

A less straightforward problem is presented by the overall total, which fell from 3,982 (*c.*14.5 per acre) in 1925 to 1,949 in 1948 and then rose slightly to 2,169 in 1966 and further to 2,319 in 1975. These aggregates conceal the extraordinary fact that the number of **House Sparrows** in the first two counts of November and December 1925 was identical (within one-third of 1 per cent, at about 2,600) and yet comfortably surpassed the

---

109. ED: In his 1925 article on the Birds of Kensington Gardens (*The Nineteenth Century and After* 98: 930), EMN gave the total number of species as 27, including Fieldfare, which took the total number of birds to 3,983. In *How Birds Live* (1927: 121) the total was given as 3,980 birds.

110. ED: The January 1949 total was also 24 but that for March 1949 was 26.

111. ED: See the annual *Reports* of the Committee on Bird Sanctuaries in Royal Parks (England) for 1928–39, published by HMSO, with later reports by the Committee on Bird Sanctuaries in the Royal Parks (England and Wales) entitled *Birds in London* (1939–47, 1948 and 1949) and *Bird Life in the Royal Parks* (1950, 1951–52, 1953–54, 1955–56, 1957–58, 1959–60, 1961–62, 1963–64, 1965–66, 1967–68, 1969–70, 1971–72, 1973, 1974, 1975 and 1976).

total for it and all other species in any later year. Something is known of the breakdown of these vast early sparrow populations, which differed greatly in different months. For instance, along the Long Water–Fountains section there were 447 in November and 636 in December 1925 (which, incidentally, exceeded the sparrow population of the entire Gardens in November 1975). A sample count in March 1926 produced a sparrow total of 980 for sections which had given a November total of 988 although, during the interim, sample counts had suggested a temporary drop for the Gardens as a whole to a level around 1,775–1,840, which was attributed to a prolonged spell of rainy weather reducing drastically the amount of food brought in by visitors to the park.

It is unfortunate that the gap before the next census, in 1948, was as long as 23 years, and that it ended with a period of abnormal conditions during and after the Second World War, which included strict food rationing. The ensuing reduction to 885 House Sparrows—a two-thirds drop, by 1,718—fortuitously went some way to fulfilling an inference made in 1925 that 'if the Park was closed for a couple of months it would not support more than about 60 of the 2,603 sparrows'. In view of the speculation over the extent and cause of this decline it is worth adding some further points from the only contemporary analysis of the previous high level. Although much careful observation had been done in preparation for the first census, the comment in December 1925 was on 'The unexpectedly vast majority of sparrows over all other species'. This is consistent with an interpretation that a heavy influx had occurred just before the census and was sustained, with one major trough, throughout the winter. There was repeated emphasis on the large amount of feeding by the public, especially with bread—'The place is a kind of vast open-air almshouse maintained by the voluntary offerings of thousands of people, and by a grant of grain from the Crown'; 'the amount of food thrown on to the ice is amazing'.

Stress was also laid on the highly uneven distribution of House Sparrows within the Gardens. In November 1925 the sparrow population was densest about the borders: the north and south Flower Walks held 425, Kensington Palace 250 plus 75 in the adjoining Dutch Garden, and the sector from the Lancaster Gate Fountains along the Long Water bank past *Peter Pan* had 447, or within a hundred of the total for the whole Gardens fifty years later. They were noted as being 'least numerous in the open or park-like parts where there are fewest people'. The ensuing month proved to be one of the coldest Novembers on record: it made no overall reduction in sparrows but it did dramatically cut the population in certain sectors. For example, in the sections west of the Broad Walk there was a drop from 595 to 238 within a month, the loss in the two Kensington Palace sections alone being 246. The back blocks of open parkland were virtually emptied.

These and other data show that the sparrow concentrations of fifty years ago were not only very large but very mobile. Although we have no knowledge of the then breeding population of the Gardens, it seems likely that the great majority of the wintering birds were bred outside, some perhaps coming in from more or less distant suburbs as autumn gleanings grew scarce in the fields: it seems improbable that the apparent surplus over summer carrying capacity can all have been drawn from the immediate built-up neighbourhood. The demonstrated capability of maintaining population during what must have been a period of heavy mortality strongly implies a continuing inflow of replacements right through until March. A similar stability at a much lower level is shown by the Cramp and Teagle 1948–49 figures (885 in December, 807 in January and 836 in March), and again by RFS in 1966–67 (642 in November, 680 in December and 544 in February).

In both 1948–49 and 1966–67, counts were kept by habitats. In both series, shrubberies and flowerbeds came out top, but much more decisively so in 1948–49 with over 80 per cent of the total, against only 31 per cent in 1966–67. Another notable feature was that, while total sparrow numbers fell from 843 to 592, numbers actually increased, often to more than double, in all habitats other than shrubberies and flowerbeds, which occupy only some 4 per cent of the entire land area.

On the basis of a Common Bird Census[112] during the 1967 breeding season, RFS estimated the number of pairs of House Sparrows at 160, which in terms of birds is about half the winter population. On the hypothesis that the summer resident population has remained fairly stable around this level, the winter immigrant population will have fallen from about 2,250 in 1925–26 to about 500 in 1948–49, about 300 in 1966–67 and about 200 in 1975–76. On that hypothesis—and it is no more—the scope for further reduction of the winter sparrow population would appear to be small, if we assume that it is unlikely to decline appreciably below the summer strength, and that the summer strength is likely to be fairly steady. It is interesting to note that, whereas in 1975 House Sparrows accounted for little over 23 per cent of the total winter population of the Gardens (against over 65 per cent in 1925), they accounted in 1967 for more than 40 per cent of the total estimated breeding population.

This somewhat extended review of the problems of House Sparrow population over the half-century seems justified in view of the interest and divergent opinions previously expressed in the literature[113] and the

---

112. ED: For a description of the Common Bird Census, see the British Trust for Ornithology's guide *Population Trends in British Breeding Birds* (J. H. Marchant et al. 1990).

113. ED: See, for example, J. D. Summers-Smith 1988, *The Sparrows*: 156–158.

improbability that an equally good opportunity will recur of combining relevant data and experience over such a long period.

A further but less serious problem is presented by changes in the numbers of **Starlings**. In November 1925 there were 411, and they were most numerous in the northern half of the interior; the carrying capacity in the absence of human feeding was then put at around 100. A month later, after a week's hard frost, the total fell by 68 to 343, this being much the largest drop for any species except the frozen-out Mallards. It was noted that the great majority were feeding on grass in sunny open parts, and there were increases except in the central and north central sections where a drop from 219 to 76 depressed the total.

A dramatic decrease was shown in the counts for 1948–49 which found only 22 starlings in December, 39 in January and 53 in March, all but a handful being on open grassland with scattered trees or under woodland.[114] A slight recovery was seen in 1966–67 with an average of 64, about 60 per cent being in the previously preferred habitats. In November 1975 there was a more marked recovery to 148 which was, however, still much below half of the 1925 level. It is possible that shifts in the relation to favoured roosts may have influenced the numbers feeding by day in the Gardens, but more frequent and active disturbance of their favoured grassland areas, and even greater compaction of the soil, should not be left out of account. The magnitude of the half-century's decline in combined numbers of Starlings and House Sparrows, from 3,014 to 692, can best be realised by pointing out that it comes to only eight fewer than the total population of all species in 1975.

Among positive trends the **Woodpigeon**, after being deliberately reduced from 241 in 1925 to 5 in 1948,[115] has since been permitted to recover to 56 in 1966[116] and 386 in 1975, giving it second place in numbers. For third place the **Black-headed Gull**, with 251, and **Mallard**, with 246, were now almost level, although in 1925 the Black-headed Gull had been almost 50 ahead, and had risen to more than double in 1948 and 1966, when the Mallard had been comparably reduced. The most outstanding

---

114. ED: W. G. Teagle has suggested (*in litt*. 1994) that the phenomenal drop in Starling numbers between 1925/26 and 1948/49 may have been at least partly due to human disturbance. All three counts in 1948/49 were made on Sundays when there was no rain. Counts he conducted at King George's Park, Wandsworth, between 16th September 1972 and 22nd September 1983 showed that Starlings were more abundant on quiet, wet days, but not exceptionally so. He suggests that 'Had the 1948/49 census been conducted on three wet weekdays it seems likely that more Starlings would have been seen, but not enough for one to doubt whether the species had suffered a great decline since the 1920s'.

115. ED: Though there were 43 on 23rd January and 53 on 13th March 1949.

116. ED: There were 138 on 17th December 1966 and 110 on 4th February 1967.

increase over the half-century, however, was that of the **Blackbird**, by just over nine times, although even so the 1975 level was hardly over 76 per cent of the 1966 peak of 250. Proportionately the **Tufted Duck** increased even more, by eleven times to 1948,[117] eighteen times to 1966 and over twenty times to 1975, this being the only other species to top the hundred mark in any of the years.

While such changes of fortune catch the eye, the series of censuses is of no less value in demonstrating the stability of the continuing core species, the arrival and consolidation of fresh colonists, and the perceptible influx of a few species which have reached peak levels nationally, as a result of a succession of mild winters or other factors. Finally, the census results on the whole confirm the success of consistently sustained policies and practices of conservation carried through by the Royal Parks staffs with the advice of the Committee on Bird Sanctuaries in the Royal Parks and the Official Observers.[118] It is, perhaps, in this intimate link between observation, study and practice that the conservation management of Kensington Gardens and other great parks of Inner London is most distinguished.

---

117. ED: In January 1949 there were 130, over eighteen times the 1925 figure.

118. ED: The Committee was abolished in 1979.

Chapter VI

Bird Protection in London

The bird-watcher who is continually in touch with the conditions under which birds live cannot help forming opinions on the steps which we take or may hope to take towards their protection. Moreover, the welfare of the London birds is a matter of general interest and for this reason I am departing once more from my first intention of keeping strictly to an account of things seen by discussing in this last chapter various points involved with bird protection. It must first of all be said that there is no great town in this country, or probably in any Christian country, where the birds are so thoroughly well protected at the present time as they are in London. They enjoy above all the influence of an almost Buddhist consideration for wild life, which is infinitely more powerful than any mere legislation against their destruction could ever be. For Britain as a whole it will be a tremendous advance if we are able during the next thirty years satisfactorily to enforce laws against the wholesale persecution of birds, more particularly rarities and birds of prey. But London has got far beyond that stage: a great part of the present generation of Londoners, if not the majority of them, are active friends of birds; food is given on such a generous scale that the bird population familiar in London depends on the bounty of man as much as on the bounty of nature. In friendship between birds and men, which is the essential part of bird protection, London has already set up a standard which the rest of England can scarcely hope to equal within the next two generations. But there are still many practical aspects which are not beyond improvement and many unexploited measures whereby more birds could be attracted.

In the first place there are the sanctuaries in the Royal Parks. These sanctuaries are not doing so well as they ought to be. The willow-wren has been lost as a breeding species in Hyde Park since they were established, although it had certainly bred there previously for at least three years in succession (1921–1923). The successful breeding of the lesser whitethroat (in 1921) a year before the formation of the Sanctuary has proved that it is not impossible for even the most specialised warblers to become summer visitors, but no fresh instance has occurred since. This comparative lack of results leads to an impression that the sanctuaries are not as attractive as they might be. I have already touched on the subject in *Birds in England*[119] where I suggested that the main need was not for more untidiness but for more suitable plants. There is an encouraging tendency to rebel against the Victorian cemetery-flora on other grounds, particularly aesthetic. The great need is for a drastic weeding out among the evergreen shrubs and their replacement by hawthorn and blackthorn, gorse, raspberries and plenty of fruit trees—cherries and apples especially. What must be borne in mind is that it is not enough simply to set aside a sequestered enclosure or two if the greater part of the vegetation consists of mere cumberers of the ground from the standpoint of the birds we hope to attract. They have two great needs: an undisturbed breeding-haven and a regular food supply close at hand upon which to rear a hungry brood. The sanctuaries have supplied only the first, and that incompletely. In the present state of Hyde Park and Kensington Gardens a colonising warbler is faced with a precarious and probably insufficient food supply owing to the insignificant extent of the undergrowth and the fact that so much of it is worthless. The successful breeding of the lesser whitethroat might be urged against this: but the records have shown it to be exceptional and the highly abnormal summer of 1921[120] may well have been not simply the only one it was, but the only one when it could have been, carried out successfully.

Birds have undoubtedly a working knowledge of the quantity of their food which is likely to be found on a given area at a given time. We do not at all appreciate the courage and foresight shown by hundreds and thousands of summer visitors in settling down and laying eggs in situations which are not at that time capable of supporting the young, but which will be capable of it by the time the young hatch. An assumption that birds have a roughly accurate method of estimating, probably unconsciously, the amount of ground which will supply enough to maintain their young is implicit in the theory of territory. If they were prone to underestimate their future needs, overcrowding and frequent starvation of young birds would follow. If, on the contrary, they overestimated, nothing would be more wasteful of energy for the

---

119. ED: *Birds in England* (1926: 218).

120. ED: This date is given in error as 1922 in EMN's *Birds in England* (1926: 216).

defending individuals or more wasteful of opportunities for the species. It seems, therefore, that they must have a good eye for possibilities and, in that case, it is significant that, of all the warblers which undoubtedly pass though these parks on their trek in search of summer territory, only two or three should think it worth remaining, and these usually fail to persuade a mate to share their choice. They look at the sanctuaries with their expert eye and their judgement is given by passing on elsewhere. Clearly there is ample cover, in the Hudson Sanctuary if not by the Long Water. Just as clearly, it seems to me, there are not enough suitable plants and shrubs to promise an adequate food supply and those that do exist are too scattered for convenience no matter where a nest is placed. To remedy this defect is by no means impossible. If the Hudson Sanctuary is dedicated to birds then that at least might be reserved entirely for vegetation favourable to bird-life, which would involve replacing a good many exotic weeds by the plants recommended by the Bird Sanctuaries Committee in its first report.[121] I should venture another suggestion which would, I believe, if carried out, improve not only the conditions for breeding birds but the amenities of the park. That is the planting of a good thick quickset hedge inside the railings separating the Serpentine Bridge–Victoria Gate road from the part of Kensington Gardens east of the Long Water. Running from the Magazine salient to the Bayswater Road, inside the Kensington Gardens railings, it would border the least frequented part of the whole open space. It would not accumulate much dirt (for the road is dustless) and it would look much better than that rather ridiculous line of low iron railings in their present nakedness. At either end it might be allowed to grow tall, into a good 'bullfinch' if possible; in the centre it would need to be no higher than the fence or the vista to the Round Pond would be lost. Such a hedge, if planted thick with hawthorn and bramble, holly, honeysuckle, wild rose and the other common hedgerow plants would be a refreshing sight in spring; it would also furnish a hunting ground for warblers within easy reach of the Hudson and Long Water Sanctuaries. On the inner side it could be accompanied by a shallow ditch. The replanting of the little enclosures on either side of the Serpentine Bridge at its east end would also be a great advantage. Other hedges could be planted

---

121. ED: See footnote on page 131.

without interfering with existing liberties, for example inside the tall railings on either side of the road from Prince's Gate almost to the Refreshment House. Even these simple measures would, I feel confident, add appreciably to the attractions of the Park and Gardens as a bird haunt. More grandiose suggestions are much simpler to make but, since the Park is not for birds alone, impracticable.

As to the shooting of crows, I have given my opinion of it in Chapter 4 (page 110). There is another bird which a good many people are itching to see slaughtered, and that is the London pigeon. They find him too much of a good thing. His numbers are undoubtedly colossal and he observes an obsolete freedom in sanitation which the City, that did precisely the same three hundred years ago, now contemplates with horror. Plagues and progress have a way of changing one's outlook: not many generations ago the Corporation protected birds which did their sanitation for them (like the kite) but, now they are so enlightened, a bird is in danger of death if it does not take enough care. Their corporate wrath has so far been restricted by the fear of being prosecuted for larceny but they are now combining with other metropolitan and local authorities to obtain Parliamentary powers to slay the offending pigeon, which meanwhile flourishes, unconscious of either the injury or coming retribution.

There is also the cat menace. In principle it is no bad thing for birds to have birds and beasts of prey to cope with. The cat is the scourge in London as weasels and stoats and others are in the country: the fact that he kills birds is no justification for smiting him. (Of course, to encourage birds in a garden with food and nest boxes and to keep a pet cat which takes advantage of their confidence to slay them is an absolutely different matter: it is one of those pieces of unconsciously refined cruelty which the kind of people who want to exterminate hawks and crows are constantly committing.) All the same, there is an implicit coincidence between the abundance of cats in central London and the curious scarcity of robins. No bird-watcher is likely to object to the elimination by a nominal tax, or some other method, of the existing population of stray cats, which do most of the damage.

The grey squirrels, undesirable aliens in this country, are not intolerable in the parks so long as they are well kept down: the present policy of preventing their increase is probably the best that could be followed both for the birds and from the standpoint of leaving no surplus population to invade other parts. Some people would like to see the corpse of the last grey squirrel in England: there is much to be said for this ambition, and the sooner this tree-rat is extirpated in country districts the better, but from the purely parochial aspect of the inner parks it might be a mistake to destroy him and have no substitute. At least there is little reason for it if he is still to flourish elsewhere.

As London spreads the inner parks become more remote from the countryside, on which the supply of many kinds of birds depends. Bird populations when they become small and isolated are not self-supporting: this is the reason why Savi's warbler and the spoonbill and osprey ceased to be British nesting birds. The home stock became scarce and insignificant, it had no reserves to draw upon and nature and man together destroyed it in detail. The same fate is now threatening the Dartford warbler, which clings in small numbers to a handful of haunts without any visible intercommunication. A small population is never secure: it appears from Hudson[122] that even the great tit was exterminated in Kensington Gardens by the frosts of the early eighteen nineties. It has recolonised since and, so long as there is a chain of other haunts linking it with the outer world, there is little fear of losing it. But if the communications are allowed to be cut a marked impoverishment of the bird-life of the central parks will inevitably take place, however attractive are the sanctuaries. Birdlovers therefore have a direct interest in the setting aside of suburban open spaces for, apart from any intrinsic merits, they feed the inner parks, which otherwise none but the adventurous would penetrate. This applies especially to robins and wrens, thrushes, tits and others, less so to migratory species which do not depend upon such links.

The conservation of sanctuaries in the outer range of suburbs should be an integral part of metropolitan town-planning schemes. Hanwell already possesses one,[123] permanently reserved by the efforts of the Selborne Society. Selsdon Wood on the North Downs is also in part to be preserved. In this case the Footpaths and

---

122. ED: W. H. Hudson 1898, *Birds in London*.

123. ED: Perivale Wood.

Commons Preservation Society and a group of local enthusiasts have taken the necessary action. Richmond Park might be classed with them: the privacy of its enclosures is traditionally undisturbed. There should be more of these. As much as possible of the valley of the Brent and Colne and the Grand Union Canal ought from any point of view to be saved as open space. But simply to take down a map and look where there ought to be sanctuaries is no more constructive than making a nature reserve of the moon: in practice one must be a tireless opportunist, supporting as vigorously as possible any projects which promise in some way to turn to the advantage of bird-life. There are places of various kinds already in existence which might without prejudice to their present development be made into strongholds of bird-life. The most important of these are the reservoirs.

By the courtesy of the Metropolitan Water Board I was enabled to make a thorough exploration of Littleton Reservoir[124] in Middlesex on 10th June 1925, three days before it opened, and again on the 13th when the ceremony had ended and the process of filling begun. The second time I walked over its great floor watching the birds which bred there; the pumps were already pouring in Thames water to fill it up. When I first knew the place it was a wood, large and comparatively unfrequented. It was unusually full of wild life but there were no extraordinary features except the immense number and variety of stinging and hunting insects. Then, a year before the War, it was all chopped down and in August 1914 they began to turn it into a reservoir. Engineering difficulties and the War delayed operations, so that by the time I saw it again in June 1925 a second chapter in the natural history of the place was approaching its conclusion. A high embankment, four miles round, encircled the site of Littleton Woods, but the middle was far from being bare waste. A great part of the bottom land had already filled with water of its own accord and the rest was overgrown with a new coat of vegetation. It was not the plants that had been in the wood but quite a fresh flora which had taken possession—Dutch and common clover, thistles, plantain, buttercups, poppies, brambles and grasses and weeds of many varieties. Three or four of the biggest and most splendid species of British dragonflies had colonised the place and the brilliant patches

---

124. ED: Now known as Queen Mary Reservoir in the modern county of Surrey.

of clover hummed incessantly with bees. But the most marvellous change was in the birds.

Down on the level floor of this man-made crater sprawled a large many-limbed lagoon several feet deep, with clumps of rushes and islets of shingle rising in places out of the water. It was a world in itself, for down at the margin the view was bounded in every direction by a sloping wall eighty feet high and nothing could be seen or heard of the universe beyond. This lagoon, in spite of its artificial and recent origin, was actually less modified by civilised man than any ancient sheet of water in England would be in this age, and it had already attracted a new and flourishing avifauna. Tufted ducks, the drakes in a majority, were present in spectacular numbers: often twenty or thirty rose in a company, with a great thumping on the water, to join another flock. They grunted freely and were far more vociferous than most seen in London. Mallard were in hundreds, the drakes again predominating; a fair handful of pochard; and plenty of coots. The rarest of these wildfowl was the shoveler: a fair party of drakes was on the water and eight of them took wing in company flying in a bunch as orderly as teal, all swerving together—an extraordinary sight within fifteen miles of London at the height of the breeding season.

Twenty-four swans, most of them probably from the Thames, were present on my first visit and at least thirty at the opening. (I was told that a black swan had remained for some time, but departed a few days before.) When I came near, one of these swans began to utter a loud petulant grunting 'Wuc-ck' or 'Wugg', which sometimes had an almost whinnying accompaniment. This was kept up incessantly for ten minutes or more; it was always monosyllabic. A few dabchicks were still diving off the inlet weir when the process of filling began but great crested grebes outnumbered them—I saw four of these take wing together, flying in line like a Japanese frieze.

Among the other birds which had taken possession of the lagoon were lapwings, snipe, redshank—these very clamorous—stock doves, one or two herons, black-headed gulls, partridges, yellow wagtails (which certainly had a nest in the long grass) and several common species. At the moment the King set the pumps working and water began to pour in there were three pied wagtails playing about the weir below the Royal stand. On the embankment whinchats, stonechats, goldfinches, linnets and tree-pipits flourished and were breeding near.

The sixty species I observed in the area of the Water Board's estate within a few hours included also such interesting birds as the red-backed shrike, tree-sparrow, lesser whitethroat, sedge-warbler, blackcap, garden warbler, long-tailed tit, yellowhammer and reed-, corn- and **cirl-buntings**. A cock of this last species sang persistently from a tree within a stone's throw of the embankment: I suspect from the date (10th June) that he must have bred or intended to breed, but there was no time to search for the nest. According to Harting's *Birds of Middlesex*[125] this has been recorded from the county only once.[126] The first **red-backed shrike** I met with clearly had a mate sitting: he was carrying a bee which he probably meant to take to her as a present but, finding himself observed, twisted it round two or three times and swallowed it with deliberation. The second was being severely mobbed by a pair of whitethroats, which evidently regarded it as a bird of prey. Separately or together they were constantly making attacks and trying to drive it out of their territory but it flew only when they plagued it and then settled again inside their boundary. When I first came up it was chasing a whitethroat, but generally the case was the other way round. It preserved all the time a Satanic calmness. All the birds, or at any rate the greater part of them, had come of their own accord to the place as soon as the works made it suitable for their requirements. It is difficult to say how many had bred: certainly a good many ducks had, for the workmen found their eggs. The Board, I believe, had given orders for them to be left undisturbed. If the shovelers bred in Middlesex the fact would be worth recording, but I could find no proof of it.

The circumstances were, of course, exceptional, and may never be repeated while London remains inhabited. An almost empty reservoir, with its tremendous bed covered by shallow lagoons and vegetation, is an infinitely finer bird-haunt than a full one, in which water, too uniformly deep for the tastes of any but proficient divers, stretches unrelieved from bank to bank. But the metropolitan

---

125. ED: J. E. Harting 1886, *The Birds of Middlesex*.

126. ED: W. E. Glegg (1935, *A History of the Birds of Middlesex*) mentions nests at Wembley Park in 1861, Hendon in 1871 and Harrrow in 1919 and 1924, which make this the fifth record.

reservoirs are, all the same, by far the greatest groups of artificial waters in this country. They might almost compete with the Shropshire meres or the Norfolk Broads in undisturbed water area. Certainly they compare very favourably with the Breckland meres about Thetford which form one of the finest haunts of rare ducks and waterbirds in England, including the gadwall, garganey, shoveler, teal, pochard, tufted duck, mallard, ringed plover, coot and great crested and little grebes, all of which breed. These meres are of almost insignificant size: of the most famous, Ringmere is only 6¾ acres, Langmere about 12 acres, Fowlmere about 18; the largest one, Didlington Lake, is about 60 acres and Thompson and Stanford Waters each about 40. The whole group consists of hardly over a dozen meres spread over an area of about twelve miles by seven and the whole number together are about a third of the size of Littleton Reservoir. The Walthamstow group alone contains, in a smaller area, twelve reservoirs, most of them forty acres or much more in extent. Altogether the Metropolitan Water Board have about 2,700 acres of storage reservoirs and Littleton, the biggest, is four miles round. The water supply of London is turning the district between Kingston and Staines into a miniature Lake District. The Board itself has stated that 'the low-lying water-bearing meadows of West Middlesex are gradually being converted into reservoirs for the ever growing needs of London and its outer suburbs ... and it looks as though the whole of this plain will be covered with a series of inland lakes'.[127] Such an opportunity of forming a stronghold of waterbirds ought not to be neglected. Properly taken it might, without much difficulty or expense, be possible to create here a paradise of water birds second only to the Broads.

The making of these reservoirs has already done much to change the bird-life of the surrounding districts. Some are already well-known to observers as the haunts of rare visitors and there are few of any size which do not hold numbers of aquatic species that otherwise would never exist in the neighbourhood. Yet the opportunities they afford for the encouragement of bird-life are still for the most part thrown away.

---

127. EMN: *Notes on Littleton Reservoir and on London's Water Supply*.

That many birds would try to breed if they were given the chance is obvious from the example of Tring, not forty miles away, which has lately added a new breeding species to the English list,[128] and others like Aldenham where coots, great crested grebes and various interesting water birds flourish in spite of all disturbance. But these are canal reservoirs and, since purity does not matter much in canal water, there is no ban on vegetation. A water supply reservoir is a different matter: even though the water is to pass through filters afterwards it must not be exposed to needless pollution. An ideal bird sanctuary would be full of aquatic plants and creatures, fringed with masses of reed and sedge, with islets and tussocks out in the water and tracts of marsh and woodland on the shore. An ideal water supply would presumably be something like a colossal swimming bath paved and margined with spotless tiles and empty of any perceptible living thing. These are extremes and the *actual* storage reservoir, with its populations of fish and wildfowl, lies between them. Naturally nothing must be done to encourage wild life which would prejudice its essential objects. All the same, much can be done to attract birds without infringing this necessary condition.

The first consideration is cover on the embankments. Waterbirds seem to have little difficulty in finding food but they cannot breed because there is no site available. The planting of embankments with gorse, junipers, trailing brambles and other native shrubs, instead of planting them with exotic evergreens as is usual at present, would give cover not only for waterfowl but for stonechats, whinchats, warblers and, particularly, goldfinches.

With the goodwill of the authorities much could undoubtedly be done towards securing for the whole year the enjoyment of several fine waterfowl at present compelled to retire elsewhere to breed. And the goodwill of the Metropolitan Water Board has already been shown in a way that all lovers of birds appreciate by the prohibition of shooting on the reservoirs, which are therefore in a sense sanctuaries already. Littleton Reservoir might without

---

128. ED: The black-necked grebe, which was first proved to breed in 1918 (C. Oldham 1919, Nesting of the Black-necked Grebe in Hertfordshire. *Bull. BOC* 39: 28–34; *Trans. Herts. Nat. Hist. Soc.* 17: 211–219). A few years later, in 1938, Tring Reservoirs also added little ringed plover to the list of British breeding species (R. C. B. Ledlie & E. G. Pedler 1938, Nesting of the Little Ringed Plover in Hertfordshire. *Brit. Birds* 32: 90–102).

much difficulty be made the greatest artificial bird-sanctuary in this country. The curse of it all is the planting of the banks in the hands of the Royal Horticultural Society, with whom Mr Harold Russell took the matter up on my suggestion some months before his death. Unfortunately, neither he nor I got much satisfaction out of them: if the melancholy example of Staines is followed, and a dreary wall of laurels and exotic evergreens is planted, the goldfinches, stonechats, whinchats, linnets, buntings and other interesting species will undoubtedly give way to a handful of the few regular wild creatures of a London shrubbery. But if native shrubs and plants—gorse and juniper, bramble, honeysuckle, wild rose and so on—the appearance of the banks will be English instead of alien and the foundation of a magnificent bird sanctuary will be assured.

THE BIRDS OF HYDE PARK AND KENSINGTON GARDENS

BY R. F. SANDERSON

A. Holte Macpherson wrote in the introduction to his 1929 list of the birds of Inner London[129] that 'Kensington Gardens and Hyde Park have never lacked regular observers'. There has, nevertheless, been no full account of the history and status of the many species of birds recorded over the years from this area. The systematic list which follows includes accepted records of 177 species by mid 1995, more than half the total yet recorded from the whole of the London Area.

In compiling this list it quickly became clear that there has seldom been a shortage of interesting birds, either in the parks or flying over, but, for periods of many years, and in contrast to the situation when Macpherson wrote, there has been a scarcity of regular observers. The systematic list should therefore be read bearing in mind that, for many species which are not resident or do not nest in the parks, the regularity of visits or passage is greater than has been reported, since the area has not always been observed daily. The most recent comprehensive records were in the period 1967–1980 when the parks were not only well covered but a Common Bird Census was also undertaken. Some of the earlier records are now mainly of historic interest. In the 1930s, for example, London was expanding rapidly and the farmland to which some species travelled out to feed was being cleared for housing. In the parks, some habitats have disappeared, such as the sheep enclosure in Kensington Gardens, the Second World War allotments and, more recently, the wildflower sanctuary—known to some as the weed garden—on Buck Hill.

There have been some spectacular losses of trees in recent years. In the 1970s several thousand elms were lost over a ten-year period to Dutch elm disease and the now infamous 'hurricane' of 15th/16th October 1987

---

129. A. H. Macpherson 1929, A List of the Birds of Inner London. *Brit. Birds* 22: 222–244. This list summarised records for the present century. Macpherson also wrote annual supplements on birds in Inner London in *British Birds* each year until 1940 when the task was carried on by Dr G. Carmichael Low (1941–45, the reports for the first and last years being undertaken jointly with Miss M. S. van Oostveen, who also undertook the year 1946 alone); subsequent reports were prepared by C. B. Ashby (1947, 1948) and W. G. Teagle (1949). Earlier, Macpherson had contributed notes on London birds to *The Zoologist* in 1889 and to *Nature Notes* in 1891 and 1892, followed by annual reports on London birds (mainly for Hyde Park and Kensington Gardens) in *Nature Notes* (from 1894 to 1908), in its successor the *Selborne Magazine* (until 1921) and then in the *London Naturalist* (1922–1927). Important summary papers on the birds of Inner London also appeared in *British Birds* (covering 1900–1950, by S. Cramp and W. G. Teagle, and 1951–1965, by S. Cramp and A. D. Tomlins).

accounted for 800 trees in Hyde Park, Kensington Gardens and St James's Park, representing up to 15 per cent of the tree cover. Most of these trees were at least 100 years old and, as such, would have exerted a considerable effect on their immediate surroundings, providing food, shelter and nesting sites. Many trees were also badly damaged, losing limbs which, particularly for horse-chestnuts, may speed up their eventual felling. Other trees, exposed by the loss of surrounding screening trees, have consequently become more vulnerable to future storms. For example, in a single storm in 1990 a further 45 trees were lost in Hyde Park and Kensington Gardens. Considerable tree planting has taken place in the last 20 years and these trees will soon be making their presence felt, but it will be decades before their fissured bark and hollowed trunks can replace fully the habitat losses of recent years.

It should be remembered that Hyde Park and Kensington Gardens are busy places. From early in the morning joggers and dogs exercise (most often around the lakes) and, by mid-morning, even the remotest parts of the parks will have been disturbed by visitors—except in bad weather and on some weekdays in winter. And then there are the open-air concerts and exhibitions in Hyde Park with which the birds somehow manage to compete. It is hardly surprising, therefore, that it is the early morning observer who is likely to see the most species in the parks. It is also most usual for visible passage across London to take place early in the day. There are several vantage points from which that can be viewed, notably from the Round Pond or the south bank of the Serpentine, but usually by 10 a.m. the peak has passed. An exception is during hard weather movements which sometimes continue all day.

A bird's-eye view of Hyde Park and Kensington Gardens is often possible from the right-hand side of an aeroplane coming in to land at Heathrow. One can imagine that to a tired or disorientated migrant the parks must appear as an oasis, an island of green surrounded by concrete. There is a green corridor, linking the River Thames at Westminster, through St James's Park, Buckingham Palace garden and the Green Park, to Hyde Park and Kensington Gardens and on to Holland Park, which is used by some birds.

During severe winter weather birds such as Redwings and Fieldfares are sometimes forced to leave their frozen feeding grounds and, finding less snow and ice in the Central Royal Parks (because the centre of London is usually a few degrees less cold than the surrounding countryside), halt their journeys for much-needed food, sometimes staying for several days.

As Max Nicholson remarked, there are opportunities in the parks for closer and more intimate views of many species, particularly wildfowl, than are possible at the most popular bird-watching sites outside the capital. The parks are an excellent place to get to know the common species and to watch the diving ducks at close quarters and observe their display rituals. Some species will take food from the hand and this tameness is referred to in the systematic list.

The parks are certainly a place in which to enjoy bird-watching, where in even an hour or so at midday, with a little practice, one can find at least 30 species. Hyde Park and Kensington Gardens are unlike St James's and Regent's Parks in not having large captive wildfowl collections, though escapes from there and elsewhere sometimes occur and some species have been introduced from time to time.

It should be noted that it is no longer possible to reassess most of the old records (e.g. of Scaup and Iceland Gull) and they have been accepted in the systematic list on the basis of contemporary standards and sources.

The principal published sources for the records included here are W. H. Hudson's *Birds in London* (1898), the papers in *British Birds* on the birds of Inner London by Macpherson and others,[130] W. E. Glegg's *A History of the Birds of Middlesex* (1935), the London Natural History Society's annual *London Bird Reports* (for the years 1936–1993, including the papers by S. Cramp & W. G. Teagle, S. Cramp and R. F. Sanderson referred to in the footnotes on page 128) and *The Birds of the London Area* (1964), and the reports of the Committee on Bird Sanctuaries in the Royal Parks.[131]

Red-throated Diver *Gavia stellata*

Rare visitor with only three records.

A slightly oiled bird, which was often observed on the bank, stayed on the Long Water from 14th March until 25th April 1934. In 1941 another bird was on the Round Pond from 27th January to 1st February, moving to the Serpentine on 1st and 2nd February. The third record was of a bird on the Serpentine and Long Water from 9th until 11th February 1948, when it was found dead, badly oiled.

Little Grebe *Tachybaptus ruficollis*

Uncommon visitor; bred last century.

There have been records in all but three of the years since 1961, usually of single birds. Favoured months have been from October to January and April, with the majority of records from the Long Water. There have been

---

130. See footnote on page 147.
131. See footnote on page 131. The Official Observers of the two parks were D. A. Boyd, Miss E. McEwen, C. H. Hawes, R. W. Hayman, A. H. Macpherson, C. H. F. Parsons, R. F. Sanderson and W. G. Teagle.

no breeding records this century but they were formerly much more common and nested on the Round Pond in at least 1867, when 20 birds and three nests were observed on 25th August. In August 1870, 25 were seen on the Round Pond and as many as 98 on another occasion. More recently, there were five present during the winter of 1979, but no more than two have been reported together since November 1980. Recent records include two birds on 14th April 1993 and one on the unusual date of 2nd July 1993.

Great Crested Grebe *Podiceps cristatus*

Breeding resident.

The Great Crested Grebe was formerly a rather scarce and erratic visitor to Hyde Park and Kensington Gardens, gradually increasing from one to two birds each year, from the first record in November 1908. It has bred regularly since 1972—the first year that breeding was recorded in Inner London—reaching a peak of six pairs in 1982 and 1983. In 1994 two pairs bred and up to six birds were on the Serpentine and Long Water in January 1995. One pair usually nests on the Serpentine island in an area now closed to boating. The nest site is quite close to the road alongside the Serpentine and affords good views of the displays and posturing of the birds throughout the breeding season.

Red-necked Grebe *Podiceps grisegena*

Rare visitor. Only five records, all involving single birds which stayed for between 6 and 11 days.

One was on the Round Pond from 31st January to 10th February 1937. An oiled bird was on the Long Water from 6th to 14th February 1950. One was recorded on 6th and 11th October 1960. Another was in Hyde Park from 22nd to 28th February 1961. The latest record is of one on the Long Water from 18th (possibly 16th) to 25th February 1979.

Slavonian Grebe *Podiceps auritus*

Rare visitor. Only five records with no particular pattern to the observations.

One was on the Round Pond from 28th November 1934 for three days. In 1937 a single was on the Serpentine from 1st to 9th February following violent easterly gales which also brought a Red-necked Grebe to the Round Pond. One, in summer plumage, was on the Serpentine on 27th April 1950 and one was on the Long Water on 11th March 1958. The most recent record is of one on the Serpentine on 26th November 1969 following overnight sleet.

Black-necked Grebe *Podiceps nigricollis*

Rare visitor with only four records.

One was on the Long Water from 3rd to 7th December 1930 and one was on the Round Pond on 28th September 1931. A bird in full breeding plumage was seen on the Round Pond, Serpentine and Long Water between 21st April and 8th May 1936. The most recent record is of one on the Round Pond on 28th February 1953.

Storm Petrel *Hydrobates pelagicus*

The only record is of one picked up exhausted near the Serpentine on 9th December 1886.

Leach's Petrel *Oceanodroma leucorhoa*

The only record, which was also the first record of a live bird in Inner London, was on 13th November 1973. The bird was seen to fly low along the Serpentine towards the Long Water, return back along the Serpentine followed by Black-headed Gulls and fly off north.

Gannet *Morus bassanus*

The only record is of an adult, associating with Mute Swans on the Round Pond, which stayed from about midday to early evening on 14th October 1952.

Cormorant *Phalacrocorax carbo*

Common visitor.

Although they can be seen throughout the year, their numbers are highest in autumn and winter.

In common with the rest of the London Area, the status of the Cormorant in the parks has undergone significant changes over recent years.[132] The first record of a truly wild bird—there were records of a few free-flying offspring from pinioned birds breeding in St James's Park in the 1930s—was in 1965, when a sick bird on the Serpentine died on 4th September. Cormorants were reported flying over Inner London for a number of years before that, and after 1966 reports came in every year of from 1 to 15 birds flying across the parks. Regular fishing forays began in 1974 and, since then, the number of birds involved and their lengths of

---

132. See P. J. Strangeman 1988. The Status of the Cormorant in the London Area. *Lond. Bird Rep.* 52: 173–194.

stay have increased. In 1994 22 birds were seen in the autumn and similar numbers were counted during January 1995.

The birds are present for most of the daylight hours and, when not fishing, they can be seen perched on the posts surrounding the Serpentine island. When the boating disturbs them they move to the posts at the north end of the Long Water. They leave shortly before dusk, flying off either towards the reservoirs to the west of London or eastwards in the direction of the growing nesting colony at Walthamstow Reservoirs.

Shag *Phalacrocorax aristotelis*

Rare visitor. Seven records involving eight birds.

One was seen on the Serpentine from 14th to 18th October 1935. One was on the Serpentine on 17th February 1937 and another was there on 17th December the same year. An immature was on the Serpentine from 27th September to 11th October 1945, when it was found dead. Two immatures were on the Serpentine on 13th March 1962 and another found in Knightsbridge was placed on the Serpentine the next day. In 1974, a sick immature in Hyde Park on 13th November which later died had been ringed on 26th July that year as a pullus on the Farne Islands. The most recent record is of a bird on 25th November 1975 which stayed for two days.

Grey Heron *Ardea cinerea*

One or two birds, occasionally more, are seen throughout the year.

There are a few records of birds feeding in Kensington Gardens as far back as 1909 and in 1929 Macpherson wrote that they often visited the Serpentine. From 1943 up to two birds were seen on various dates each year. By 1965 there were up to three birds present in autumn and winter. Since 1969 one or two birds have been seen throughout the year and individuals have been seen to roost on the west bank of the Long Water.

Herons have nested in Regent's Park since 1968 and in Battersea Park since 1990. More recently, a few wooden platforms have been fixed in trees on both banks of the Long Water to encourage the birds to start a heronry in Kensington Gardens. The herons are sometimes harried by Carrion Crows and make their escape by sheltering under the trees on the east bank.

In March 1995 a woman threw whole chicken breasts to two herons which readily consumed them. She said that she had been feeding them for some two years. K. C. Osborne (*in litt.* 1995) notes that the herons can be very tame. He has seen one grab bread thrown to the ducks and pigeons on the west bank of the Long Water.

Mute Swan *Cygnus olor*

Breeding resident.

In Kensington Gardens they have had mixed fortunes. In 1937 there were over 10 pairs on the Long Water and Round Pond and in early 1938 there were over 30 birds on the Round Pond. Nesting is known to have taken place in 1947 and subsequently in most years, reaching a peak of four pairs in 1955. Unfortunately, egg stealing was a problem so their success level was low and there was no nesting in the Gardens between 1956 and 1963. During the early 1950s maximum autumn counts varied between 42 in 1950 and 63 in 1955 but thereafter wintering numbers fell to maxima of nine or fewer. An upturn began in the 1970s with regular breeding again in Hyde Park and Kensington Gardens and higher wintering numbers, with up to 29 in 1973. In 1994 two pairs bred and 38 birds were present on 7th December. In both 1994 and 1995 one pair, nicknamed William and Mary, frequently entertained visitors to Kensington Gardens by escorting their cygnets between the Long Water and the Round Pond. This family could sometimes be found just sitting on the grass midway between the two areas of water, in one of the few quiet areas of the park.

An extensive ringing programme is taking place in the Thames Valley. Several ringed birds have appeared on the Round Pond. The Mute Swans wintering there in 1993/94 included birds ringed at Hampton, Staines, Walton-on-Thames, Twickenham, Molesey, Henley-on-Thames and Bushy Park. Mute Swans ringed as cygnets in Hyde Park and Kensington Gardens have been reported at Hampton and Hendon.

Although lead poisoning from fishing tackle is almost a thing of the past, other problems still beset the Mute Swans. In Hyde Park in 1994 two died from botulism, one was injured by a dog and another was tangled in kite wire.

Bewick's Swan *Cygnus columbianus*

Rare visitor with only one record in the parks, of a bird on the Serpentine in fog on 6th March 1948.

There have also been a few reports of birds flying over. In 1974, a flock of 13 swans flying west on 31st October were thought by the observer to be Bewick's. In 1979, separate flocks of 40 and 80 Bewick's flew north-east on 23rd February.

Whooper Swan *Cygnus cygnus*

The only record is of one flying north-west over Hyde Park on 29th November 1978.

White-fronted Goose *Anser albifrons*

The only definite record is of about 200 flying north-east on 23rd February 1979 but a further 80–100 grey geese flying that way the same day were probably also Whitefronts and 16 which flew over north-east on 22nd February 1972 were considered by the observer probably to be Whitefronts.

Greylag Goose *Anser anser*

A feral population established from pinioned collections.

In the 1980s Greylags were comparatively infrequent in Hyde Park and Kensington Gardens. Numbers have since increased and as many as 44 were present on 18th December 1990 and 68 on 6th March 1994. However, numbers present vary according to two main factors: the time of year—most birds leave the parks during the breeding season—and the time of day. Ringing has established that the birds from the main flock in St James's Park spend much time grazing there on the lawns until they become crowded with people. The birds then disperse to Buckingham Palace garden, Hyde Park and Kensington Gardens. Co-ordinated counts of the four areas on 22nd February 1995 produced a total of 97 geese.

Breeding is controlled within the central parks, but some geese leave the parks to nest and moult, returning with young in the autumn.

Canada Goose *Branta canadensis*

A feral population established from introduced birds and supplemented by additions from outlying feral populations.

Full-winged birds were released in Hyde Park in July and December 1955 and 16 pinioned birds were introduced there in 1962, since when breeding took place unchecked every year until the mid 1980s, when egg pricking began in an effort to control numbers.

The highest count was on 31st January 1987, when 553 were present in Hyde Park and Kensington Gardens. The peak count in 1993 was 443 in August. In Greater London and Middlesex as a whole numbers of Canada Geese were three times higher in 1991 than in 1983, an increase of about 16 per cent p.a.[133] The increase in Hyde Park and Kensington Gardens actually took place before that period and numbers may now have stabilised as a result of egg pricking. In 1994, 115 eggs were pricked from the first laying, and a further 77 from subsequent laying, in 28 nests in Hyde Park.

---

133. H. Baker 1992. Status of the Canada Goose, the Greylag Goose and other introduced geese in Greater London and Middlesex, 1991. *Lond. Bird Rep.* 56: 175–182.

A few of these geese can be seen flying in to roost in Hyde Park: they can be watched from the Serpentine bridge at dusk.

An extensive ringing programme took place during the moult in 1994 throughout Greater London, when more than 1,700 Canada Geese were ringed. Early indications confirm a considerable amount of movement to and from Hyde Park and Kensington Gardens. In the first 12 months after the ringing programme began geese had been observed there which had been ringed in Battersea Park (60 ringed individuals), Wandsworth Common (18), Clapham Common (2), Regent's Park (2), Walthamstow (1), Barn Elms (4), Hampton (2), Ravenscourt Park (1) and Osterley Park (1). Of the 84 birds ringed in Hyde Park, only some 60 per cent have been recorded there subsequently. It will be interesting to discover how widely the others have dispersed.

[Brent Goose *Branta bernicla*]

A record of one with Canada Geese at the Round Pond on 18th February 1989 is treated as relating to an escape.

Shelduck *Tadorna tadorna*

Rare visitor with six records of birds treated as of possible wild origin.

One was on the Serpentine on 19th March 1943. One flew over the Serpentine on 3rd July 1967 and a pair (possibly from Regent's Park) was present on 8th April and 22nd May 1969. The three most recent reports are of one on the Serpentine on 19th February 1979, one in Kensington Gardens on 9th January 1984 and about 12 on the Round Pond on 19th January 1985, where 'a few' had been seen the week before during a period of cold weather. There are now a few free-flying feral birds in St James's Park and Buckingham Palace garden.

[Mandarin Duck *Aix galericulata*]

One on the Round Pond on 9th January 1940 was thought at the time to have come from the well-established feral population in the wooded country around Virginia Water and Windsor Great Park, though it seems more likely that it was an escape from a local wildfowl collection.

Wigeon *Anas penelope*

Rare visitor.

There are a few free-flying birds arising from pinioned collections so records of this species are apt to be dismissed as relating to escapes. There was even a doubt about the first record, of a pair on 11th February 1922, 'believed to be wild'. Since 1940 there have been reports in nine years, six of the records falling between November and January. The most recent record of a bird of assumed wild origin is of a drake grazing by the Long Water on 7th January 1954. A pair of pinioned birds was introduced to the Long Water in 1966, the drake surviving there for a couple of years.

Gadwall *Anas strepera*

Rare visitor.

A very tame male was a regular winter visitor to Hyde Park and Kensington Gardens from the winter of 1920/21 until 1932/33, but stayed throughout 1927 and 1928.[134] For the first few winters he hardly ever left the Round Pond and was rarely to be seen more than a few yards from its western bank. This bird often paired with a Mallard and there were thought to have been offspring. A second male was seen with him for a short time in autumn 1922. Subsequent records are of a female on the Round Pond on 18th March 1939, one in eclipse on the Round Pond on 14th July 1941, one on 1st January 1949, a drake on the Round Pond on 18th November 1950 and a drake on the Serpentine on 1st March 1951; a pair on the Long Water on 13th April 1940 had probably come from St James's Park. The only recent record is of one in Kensington Gardens on 13th January 1992.

Teal *Anas crecca*

Rare visitor.

In 1929 Macpherson reported that Teal, which from their behaviour were clearly wild, sometimes appeared on the Serpentine but did not stay long. There have been 18 records from the parks since 1944, including two reports of birds flying over. Mainly single birds were reported, between September and November, in 1944, 1948, 1952 (two birds), 1953, 1967 and 1970. In 1947, there were three drakes on 13th June and singles on 9th September and 27th October. A pair was present on 28th February and 5th March 1951. Two flew over at dawn on 17th April 1951 and three flew low over the Round Pond on 25th November 1948. In 1971, a pair was on the Round Pond on 11th February and 30th March and two birds were on the Long Water on 20th August. Since then the only records are of two on the Serpentine on 24th April 1980 and a drake on the Long Water on 29th November 1987.

Mallard *Anas platyrhynchos*

Common resident.

---

134. As mentioned in the footnote on page 6, this bird may not have been wild.

We are fortunate to have regular annual counts stretching back to the 1950s. The highest-ever count was 608, including 332 drakes, in December 1956. Since then there has been a dramatic decline, as indicated by maximum counts at intervals:

1965	523 (363 drakes)	(March)
1970	465	(September)
1975	420	(Jan–Mar)
1980	350	(Jan–Mar)
1985	132	(April)
1988	178	(January)
1994	122 (90 drakes)	(March)

This decline is much more marked than the overall decrease in counts across London, as reported in the *London Bird Report*.

There has also been a decline in the number of breeding birds, though these have not been reported as regularly as the maximum count:

1965	77 young
1968	94 young (32 broods)
1975	30 broods
1979	20 broods
1980	52 broods
1982	29 broods
1983	12 broods
1994	32 young (8 broods)

Macpherson reported that the majority of nests in Kensington Gardens were in trees, often some distance from water.

Botulism on the Serpentine in 1994 accounted for the deaths of 81 Mallard and by 6th November there were only 23 in the two parks. The total had risen to 51 by 25th January 1995. The reasons for the long-term declines in wintering and breeding numbers are a mystery.

Pintail *Anas acuta*

Rare visitor with only four records, involving six birds, all on the Round Pond.

There was a pair on 22nd April 1936, a drake on 3rd and 9th March 1942, a drake on 24th December 1946 (when the Pond was almost completely frozen) and a pair on 8th September 1976. Other records are assumed to relate to escapes.

Garganey *Anas querquedula*

Rare visitor. Three records involving six birds.

There were two drakes and a duck on the Serpentine on 12th March 1952 and, in 1968, two on 19th July and one on 10th September, all on the Round Pond.

Shoveler *Anas clypeata*

Mainly a winter visitor.

There have been reports in almost every year since 1967, with only isolated records before that since the first report in 1939. There has been a steady increase in numbers of Shoveler and in their length of stay. From two individuals in 1974 there were successively higher peak counts of 44 on 18th January 1981, 50 on 20th December 1985 and 68 on 11th January 1986. Since then numbers have been variable. There were 40 in January 1993 with 25 still there in March.

The birds favour the north end of the Long Water, spending much of the day close to either bank under cover of the overhanging trees and shrubs. Some of that cover was lost in 1994 when two old horse-chestnut trees fell and some bank repair work was begun.

A drake ringed in Hyde Park on 1st March 1986 was shot in the Pinega Region, near Arkhangelsk in northern European Russia—2,959 km distant—on 15th May 1987.

During the icy winter of 1986 several of the Shoveler became sufficiently tame and hungry to take bread thrown by visitors.

[Red-crested Pochard *Netta rufina*]

Free-flying birds from the collections in the other Royal Parks probably account for all records of this species in Hyde Park and Kensington Gardens. As many as 12 were on the Round Pond on 17th February 1953.

Pochard *Aythya ferina*

Common resident and winter visitor.

A flock of between 40 and 50 on the Serpentine on 4th February 1904 was an unusually large gathering at that date. In 1929 Macpherson reported that during the last few winters they had frequented the Round Pond, where they had become remarkably tame; they arrived in autumn and left in March. During cold weather in February 1929 there were over 50 on the Serpentine; they left for a time when the water became entirely frozen but returned later in diminished numbers. Since then wintering numbers have varied mostly in the range 20 to 70, with a maximum report of 113 on 21st January 1971.

Breeding began in 1959 and young have been seen in most years since. There were eight broods in 1971 but three or four broods were more usual in the early 1990s.

A drake ringed in Kensington Gardens on 1st December 1977 was shot at Tyumen, in western Siberian Russia—4,325 km distant—on 10th May 1979.

Some birds will come for food thrown by visitors, but frequently in winter small numbers remain remote from the others, preferring to spend the day roosting under cover of the trees by the Long Water.

[Ferruginous Duck *Aythya nyroca*]

Reports from the Round Pond in the early 1860s are not accepted as fully authenticated. Subsequent records, such as one on the Round Pond on 24th January 1938, are treated as relating to escapes.

Tufted Duck *Aythya fuligula*

Common resident and winter visitor. Wintering numbers range widely with influxes in severe weather.

Prior to 1900 the Tufted Duck was a scarce winter visitor. There were over 250 on the Serpentine at the end of January and beginning of February 1928 and during cold weather in February 1929 there were about 250 on the Serpentine; they left for a time when the water became entirely frozen but returned later in diminished numbers. In the early weeks of 1937 there were generally over 300 on the Round Pond and there were 307 (223 males) on the Long Water and Round Pond on 18th December that year. A few counts in the 1950s were of 200–240 on the Round Pond in February and March. By the 1960s maxima of 300–380 were being reported, rising to a record 664 on 1st January 1970. There then followed a long period with no published winter counts until 1986 when, on 11th January, only 164 were present. In 1991 wintering numbers in the two parks were around 100 and in 1992 there was a maximum count of 70 on 3rd February. On 25th January 1995 there were 144, 18 of which were on the Round Pond.

At least one brood was seen in Hyde Park before the First World War and the species nested regularly in the parks from 1924 until 1938 but then not until 1954, since when it has nested regularly, with a few pairs nesting each year. In 1994 there were four broods. They nest quite late, with young often not hatching until mid July or August.

Birds ringed in Hyde Park and Kensington Gardens between 1975 and 1993 have been recovered overseas from Russia (5, including 3 from the west Siberian plain in Asia, over 4,000 km distant), Finland (5), Denmark (4), Sweden (3), Netherlands (3), France (3), Norway (2), Germany (1) and Northern Ireland (1). One bird was retrapped in Hyde Park 14 years after it was first ringed.

Tufted Ducks will come readily for food thrown by visitors, although in recent years they have tended to be crowded-out by the increased number of Canada Geese.

Scaup *Aythya marila*

Rare visitor, formerly more frequent.

In recent years there have been very few records but back in the 1920s, 1930s and 1940s one or two Scaup were seen almost every year following the first record in autumn 1923. One was on the Round Pond or Serpentine for most of January 1960. The following year one was seen twice in October. The next records were of a pair on the Long Water from 5th to 8th April 1971 and, most recently, an immature male in Hyde Park on 27th February 1980.

Long-tailed Duck *Clangula hyemalis*

Only one record, of a duck on the Round Pond which seemed quite tame on 19th and 20th February 1952.

Common Scoter *Melanitta nigra*

Rare visitor. Three records involving four birds.

Two were seen on the Long Water on 13th April 1928. A drake there on 2nd July 1940 may have been oiled. Finally, one was on the Serpentine on 30th April 1979.

Goldeneye *Bucephala clangula*

Rare visitor with nine records this century.

An immature male was on the Round Pond from 19th October 1934 until 18th May 1935, when it was apparently drowned by Mallards; during its stay it made a short visit to the Serpentine. An immature male was on the Round Pond from 15th October to 5th November 1947 and again between 6th January and 15th April 1948; it also visited St James's Park lake. An adult male was on the Round Pond on 15th March 1950. Singles were present on 18th December 1972 and 25th November 1975. The most recent are all winter records, with two birds on the Serpentine from 10th to

12th January 1982, three brown-heads on the Serpentine on 16th January and one on 13th December 1986, and a drake on the Round Pond on 7th January 1993.

Smew *Mergus albellus*

Rare winter visitor with records in 10 of the years since the first in 1928.

One was on the Serpentine on 7th and 8th February 1928. During a cold spell one was on the Long Water on 20th December 1938; it moved to the Serpentine the next day where it was joined by a second bird from 23rd and by a third from 24th, all leaving when the thaw came soon after. One was on the Round Pond on 17th January 1939. A redhead was in Kensington Gardens on 3rd January 1948. One was on the Serpentine on six dates between 28th January and 15th February 1952. A redhead was on the Long Water on 31st December 1957. The most recent records are of a pair in Hyde Park from 18th to 22nd January and 5th February 1960; a drake on the Serpentine on 10th January 1962; single redheads in Kensington Gardens on 5th January 1963 and on the Serpentine on 14th January 1966; and singles on the Long Water on 1st January and the Round Pond on 8th March 1979.

Red-breasted Merganser *Mergus serrator*

Rare visitor with three records involving five birds, all on the Serpentine.

There was an adult male on 12th February 1922, three on 1st February 1976, one staying until 9th, and a drake from 19th to 23rd February 1979.

Goosander *Mergus merganser*

Rare visitor with seven records, all between December and March.

One on the Serpentine stayed for most of the morning of 12th January 1928 and a brown-head was there on 14th December 1933. On 9th March 1948 a brown-head flew over the Long Water. There were then three records in three years, with a single bird on the Round Pond on 5th February 1972, two in Hyde Park on 3rd December 1973 and one flying south over Hyde Park on 4th December 1974. The only recent record is of a female on the Serpentine on 17th February 1988.

Red Kite *Milvus milvus*

Glegg notes that, writing in 1777, Thomas Pennant mentioned two instances of this former common scavenger of the streets of London breeding in Hyde Park.

harrier *Circus* **sp.**

A ring-tail, probably a Hen Harrier *C. cyaneus*, was seen over Kensington Gardens at 7.10 a.m. on 6th November 1974.

Sparrowhawk *Accipiter nisus*

Rare visitor but might be expected to become more common since now breeding in Regent's Park and Battersea Park and widely reported in Inner London in the breeding season.

There are a few early records, since the last century, mostly without specific dates being given. In 1953 there were two records in Kensington Gardens, on 7th May and 1st June, probably associated with a pair which attempted to nest in Holland Park that year. It was 1975 before the next sighting was reported—one over Kensington Gardens on 2nd November. There were singles on 25th April 1977 and 31st December 1978, also in Kensington Gardens. The next sightings were of one on 16th April 1990, two circling on 18th May 1993 and one on 7th July 1993. In 1994 there was one on 19th January and another on 15th November which was harried by Carrion Crows for several minutes.

Common Buzzard *Buteo buteo*

Surprisingly, the only record is of one which flew over Hyde Park in early October 1926. Buzzards have not infrequently been noted over Regent's Park. Many reports of buzzards flying over Inner London have lacked sufficient detail for the records to be authenticated.

Osprey *Pandion haliaetus*

The only accepted record is of one which attempted to fish by the Serpentine bridge on 2nd May 1967. An earlier report, of one flying south over Hyde Park being mobbed on 27th September 1965, may have been authentic but was not accepted at the time.

Kestrel *Falco tinnunculus*

Uncommon resident.

A pair or two sometimes nest just outside the parks—on buildings to the north and south—and birds are seen displaying and making feeding sorties into them, but there has never been proof of breeding within the parks. During the Common Bird Census in 1994 there were only two records—one on 19th April and two together on 16th July—but one or two are likely to be seen throughout the year.

During the 1960s one bird had a strategic perch on a tree inside the lido. It would suddenly swoop on an unsuspecting feeding flock of House Sparrows, sometimes successfully.

Merlin *Falco columbarius*

Rare visitor with only three records.

One was in Kensington Gardens on 24th August 1936 and another was watched flying low between elms there on 22nd August 1943. On 12th October 1973 one was watched flying around the Long Water. It rose up as if to take a bird in flight and was seen to have something hanging from its talons. The observer was unsure whether this was prey or jesses, so the possibility of the bird being a falconer's escape cannot be ruled out.

Hobby *Falco subbuteo*

Surprisingly, given the frequency of observations from Regent's Park, the first record for the parks was of one in Hyde Park as recently as 9th September 1994.

Peregrine *Falco peregrinus*

Only one published record, of one high over the Serpentine on 18th November 1944, but E. M. Nicholson (*in litt.* 1995) clearly remembers seeing one over Hyde Park Corner in the late 1940s and two were seen flying over the adjacent Brompton Road, Kensington, on 26th February 1922. Inner London records are not all that infrequent.

Red-legged Partridge *Alectoris rufa*

Two records of singles, on 27th March 1973 and 7th May 1986, both in Kensington Gardens.

Grey Partridge *Perdix perdix*

A young and very tired bird picked up in Kensington Gardens a few days before 4th February 1922 and a bird by the Round Pond on 3rd May 1972 are the only records.

Pheasant *Phasianus colchicus*

One in the Orangery area of Kensington Gardens on 30th April 1952 is the only record.

Water Rail *Rallus aquaticus*

Writing in 1935, Glegg noted that it had been recorded in winter from the Serpentine during the present century. There are no later records.

Corncrake *Crex crex*

One was found dead, floating on the Serpentine, on 6th October 1913. This is the only record.

Moorhen *Gallinula chloropus*

Common resident.

In 1929 Macpherson described it as resident in considerable numbers, breeding freely, as it had been when he wrote in 1891. Recent records of regular breeding stretch back to at least 1948, with an estimated eight pairs nesting in 1970. They frequent the Dell in Hyde Park and the banks of the Long Water, with usually two to four pairs nesting. There used to be plant tubs among the fountains in the Italian Gardens which were favoured nest sites in the 1950s and 1960s.

Numbers increase in winter, with a peak count of 43 in December 1967. There have been fewer in recent years: 21 on 25th January 1995 is now a more typical count.

Coot *Fulica atra*

Common resident.

Bred in Kensington Gardens in 1937 and in Hyde Park in 1938, then there was a long gap in breeding records until 1953 and 1954 with regular breeding from 1957. Nests can be seen along either bank of the Long Water and on artificial rafts on the Round Pond. There were eight nests in 1994.

In 1929 Macpherson noted that in hard weather flocks of a dozen or more Coots often appeared on the Serpentine. Numbers still increase in hard weather and in February 1963 there were 170 present. By contrast, on a mild 25th January 1995 there were only 46.

Oystercatcher *Haematopus ostralegus*

E. M. Nicholson heard and saw one flying over Hyde Park, near Marble Arch, at about 4 p.m. BST on 9th September 1934. It was flying fairly low and as if lost; eventually it went off in the direction of Lancaster Gate. This is the only record.

Avocet *Recurvirostra avosetta*

Two records involving five birds.

On 13th March 1970 three landed briefly by the Serpentine and on 24th May 1992 two circled the Round Pond at 6.10 a.m.

Ringed Plover *Charadrius hiaticula*

Rare visitor with only two records.

One was by the Serpentine in April 1932 and one by the Round Pond on 26th September 1940.

Golden Plover *Pluvialus apricaria*

The only record is of five which flew westwards with Lapwing on 30th January 1976.

Lapwing *Vanellus vanellus*

Passage migrant.

Lapwing can be seen crossing London most years in June and July as they return from nesting sites and during the winter months as they move from frozen ground and search for new feeding areas. In 1970 the June passage was noted over Hyde Park and Kensington Gardens on 13 days from mid month and in December hard-weather movements were noted from Christmas Day to the end of the year. Numbers of birds vary from a few small flocks in summer to over 800 on 31st December 1978.

There have been only two reports of Lapwing landing in the parks since 1950. One was on the diggings for the underground car park in Hyde Park on 19th March 1969 and another the following year by the Round Pond on 15th March.

Sanderling *Calidris alba*

The only accepted record is of one at the edge of the Round Pond during a snowstorm on 15th February 1979. Macpherson did not accept a report of one said to have been seen by the Serpentine on 30th July 1900.

Little Stint *Calidris minuta*

The only record is of one by the Round Pond on 14th May 1976.

Dunlin *Calidris alpina*

Rare visitor.

There have been 17 records, spanning all months except January, February and June. The most recent reports are as follows: one by the Round Pond on 23rd March 1971; two flying over Kensington Gardens on 27th October 1971; singles by the Serpentine on 25th October and by the Round Pond on 10th November, both in 1975; and one by the Round Pond on 24th and 25th September 1985.

Common Snipe *Gallinago gallinago*

Rare visitor with 18 records.

Snipe were reported in almost every year during the 1970s when the parks were being well watched in the early morning. There was no particular pattern to the occurrences, except for an absence in the summer months. Recent records include singles on 19th February and 22nd October 1979, one flying over on 22nd April 1980 and one on 7th January 1985.

Woodcock *Scolopax rusticola*

Rare visitor with 18 records.

Half the records refer to the 1970s when the parks were well watched. March and November are the most frequent months of observation. Birds were seen flying over in December 1967, June 1968 and December 1970. An injured bird was found on 26th November 1971 and another had flown into a building on 16th March 1977. Singles were in Hyde Park on 13th July 1978 and 3rd December 1981 and in Kensington Gardens on 3rd April 1981; these are the most recent records.

At least three of the records are of injured birds; it is not that unusual for migrating birds of this species to fly into obstacles en route.

Bar-tailed Godwit *Limosa lapponica*

The only record is of 40 watched flying north-east at 7.15 a.m. on 23rd April 1974.

Whimbrel *Numenius phaeopus*

Rare passage migrant with five records of one or two birds flying over.

One flew over Kensington Gardens on 28th September 1934. Two flew north-north-west on 25th April 1968, one flew north-west on 5th August 1970 and two flew south-west on 22nd August 1980. Most recently, on 20th May 1982 one was seen flying north-west over Hyde Park.

Curlew *Numenius arquata*

Rare passage migrant.

There have been eight records of one to three birds flying over, plus an exceptional record of 12 on 28th April 1959. At 8 a.m. on 20th December 1938 one flew, calling, over the whole length of the Serpentine and then disappeared westwards. One flew round the Round Pond, calling, on 8th November 1941 during overhead fog. One flew east on 16th March 1970, one flew over calling on 18th April 1975 and singles flew over on 8th April and 5th August 1977. In 1980, two flew west on 22nd July and three flew west on 22nd September. The most recent record is of one in Kensington Gardens on 26th January 1992.

Redshank *Tringa totanus*

Rare visitor with nine records, including five of birds overflying the parks.

Several were heard calling over the Serpentine on the evening of 24th November 1904. There were singles in Hyde Park on 1st July 1921 and 13th May 1927. One was seen and heard flying over Kensington Gardens on 2nd August 1929. One was seen flying up the Serpentine on 8th October 1934 and another was by the Round Pond on 20th August 1939. Five waders flying south over the Round Pond on 12th March 1947 were 'probably' Redshank. One flew south on 5th August 1970 and one was in Kensington Gardens on 30th November 1978.

Greenshank *Tringa nebularia*

Three reports, all of birds flying over.

One over the Round Pond on 18th September 1951 was the first for Inner London. One flew south-west on 26th June 1973 and one flew westwards on 7th July 1975.

Green Sandpiper *Tringa ochropus*

Two records involving three birds.

Two flew over Hyde Park at dusk on 3rd October 1947 and one was at the eastern end of the Serpentine on 16th November 1960.

Common Sandpiper *Actitis hypoleucos*

Passage migrant.

In 1929 Macpherson noted that it was observed fairly frequently on spring passage by the water in Kensington Gardens and Hyde Park, but it was not noticed as often during the autumn migration. Several probably visit the parks each year. In the 1970s, when the parks were watched regularly, there were between 2 and 18 'bird days' during the spring and autumn migrations, with five birds together on 17th August 1970 and 31st

July 1980. Recent records include one on 23rd April 1992 by the Round Pond, one on 13th May 1994 and two on 15th July 1994 flying around the Long Water.

Arctic Skua *Stercorarius parasiticus*

Rare passage migrant with only two records.

One flew low over Hyde Park on 16th May 1916 and a light-phase bird was over the Round Pond on 29th October 1974.

Little Gull *Larus minutus*

The only record is of a freshly dead immature bird found in a waste bin by the Serpentine on 15th January 1960. When this bird was taken to the Natural History Museum the staff were more interested in the live mites around its bill than in the bird itself!

Black-headed Gull *Larus ridibundus*

Abundant autumn and winter visitor.

In 1929 Macpherson noted that very few of the Black-headed Gulls which frequented the parks spent the night there. Nearly all left just before sunset and repaired to the Thames and the reservoirs in its neighbourhood. They generally departed in March. Juveniles were rarely seen in the parks until they had effected the change to first-winter plumage.

There are reports of Black-headed Gulls taking food from the hand in urban parks at least as long ago as 1898, but there do not seem to be any reliable counts until 1925, when the series of bird censuses in Kensington Gardens began. The counts of 3rd and 5th December 1925 show 440 in Kensington Gardens. On 12th November 1950 there were about 400 on the Round Pond and 440 at the Long Water and Serpentine. A count on 4th December 1994 revealed 300 in Kensington Gardens plus a further 130 in Hyde Park. On 25th January 1995 the figures were 470 and 58, respectively. Counts vary, with influxes of gulls if food gets scarce elsewhere: 1,500 were counted in the two parks on 27th February 1968 and 1,080 on 11th February 1989.

The birds leave at dusk, flying off towards the west London reservoirs. First arrivals can be expected in early July, some staying until the end of March.

Birds ringed in Hyde Park and Kensington Gardens between 1975 and 1994 have been recovered overseas in the breeding season from the Netherlands (4), Sweden (4), Denmark (2), Finland (2), Germany (2), Russia (2), and Belgium, Poland, Lithuania and Norway (1 each). One ringed in Hyde Park in January 1978 was retrapped at Hammersmith 14 years later.

Common Gull *Larus canus*

Common winter visitor but immature birds can occasionally be seen throughout the year.

Never as numerous as the Black-headed Gull, the highest reported total, for both parks combined, being 84 on 29th November 1967. The 1925 census in Kensington Gardens showed 14 on 3rd and 5th December. On 4th December 1994 there were 16, plus a further 14 in Hyde Park. Like the Black-headed Gull, therefore, the Common Gull has shown little change in numbers over the last 70 years.

The parasitic habit involving several Common Gulls chasing sometimes just one unfortunate Black-headed Gull all over the sky until it drops its beakful of food can be seen at quite close quarters by the Round Pond. Just occasionally one or two will take food from the hand, but it is not common practice.

Lesser Black-backed Gull *Larus fuscus*

Resident.

In 1929 Macpherson noted that they rarely visited the parks. In the 1930s and 1940s one or two birds were reported on specific dates, mostly in autumn and winter. By the 1950s one or two were staying throughout autumn and winter, the highest count being of four birds. The species has been reported throughout the year since 1968, with a maximum of 10 on 29th March 1993. The *London Bird Report* for 1993 comments that the Inner London breeding population is probably increasing. It seems likely that some may be nesting on buildings in the vicinity of Hyde Park and Kensington Gardens.

There are mixed feelings towards the increase in numbers of this species, since it is one of the chief suspects for the disappearing ducklings. In the breeding season one or two are often found in the vicinity of the young Mallard and Tufted Duck and have been seen to take some of them. On 24th May 1995 there were seven on the Long Water.

Birds of the nominate race from Scandinavia have frequently been reported.

Herring Gull *Larus argentatus*

Regular visitor.

In 1929 Macpherson described this species as often seen on the park waters in winter. There were exceptional records of up to 60 on the Round Pond in January and February 1948, 23 in December 1949 and 21 in January and December 1950. There were 12 on 20th February 1995.

Since 1968 there have been records of birds throughout the year, almost certainly coming from St James's Park or Regent's Park in both of which the species nests. There were unsuccessful attempts at nesting on the Serpentine island in 1974 and 1978.

Birds of this species are, like the previous one, quite active when recently-hatched ducklings are on the lakes.

Iceland Gull *Larus glaucoides*

An adult stayed on the Round Pond and Long Water from 20th February until 10th March 1942, killing two feral pigeons during its stay. An immature was also reported at the Round Pond on 8th March that year.

Glaucous Gull *Larus hyperboreus*

An adult on the Round Pond on Christmas Day 1941 was one of two birds in the London Area that winter. These sightings were associated with unusual numbers of Glaucous Gulls reported from six counties down the east coast from Shetland to Kent.

Great Black-backed Gull *Larus marinus*

Rare visitor.

The species was apparently less scarce in the past, with 18 records during the 1940s and 1950s all, except one record of three, of single birds. There were seven records in 1968, three in 1969 and two in 1970. All were of one or two birds, between September and April. The next record was 25 years later, with one on the Serpentine on 25th January 1995.

Kittiwake *Rissa tridactyla*

In 1929 Macpherson noted that it was over 30 years since he had observed it on the Serpentine. The only record this century is of one over the Serpentine on 24th February 1946, following gales the previous day.

Sandwich Tern *Sterna sandvicensis*

Four isolated records, all in September.

There was one in Hyde Park on 2nd September 1908 and one at the Round Pond on 11th September 1937. The two most recent records are of

15 over the Serpentine after heavy rain on 20th September 1967 and one there on 5th September 1973.

Common Tern *Sterna hirundo*

Rare passage migrant. Some observers have not distinguished between this species and Arctic Tern *Sterna paradisaea*, so many records refer to 'Commic' Terns. There are no definite records of Arctic Tern.

The most likely months to see these birds have been August and September, with sightings in most years during the 1970s when the parks were well watched. There have been only four spring records this century. Numbers vary from ones and twos to a high of 21 on 27th September 1969 but Macpherson saw a flock of about 20 terns, which he believed to be Common, flying fairly high over the Serpentine in an easterly direction on 12th May 1906, which was a hot sultry day. The most recent report is of one in Hyde Park on 21st August 1988.

Little Tern *Sterna albifrons*

One on the Round Pond in September 1932 is the only record.

Black Tern *Chlidonias niger*

Rare passage migrant.

One was seen over the Serpentine on 27th May 1935 at a time of marked passage through the Thames Valley. The only other spring records are of one over the Serpentine on 21st April 1969 and a single on 5th May 1974.

The only autumn observations were in 1981, when one over the Round Pond on 23rd September was joined by a second on the 26th; both birds were seen to depart high to the south early on the 28th.

Guillemot *Uria aalge*

The only record is of an oiled bird on the Serpentine on 3rd January 1972 which was taken to St James's Park for treatment and later released.

Razorbill *Alca torda*

The only record is of an immature on the Round Pond from 6th to 18th October 1948. It was frequently watched catching small fish and occasionally was seen on the bank

Little Auk *Alle alle*

Rare visitor. Two records.

One was picked up alive on the path by the Round Pond on 31st December 1929; it was identified at the Natural History Museum. The only other record is of one on the Long Water at 2.20 p.m. on 30th October 1992 which flew to the Serpentine, leaving the park 30 minutes later.

Feral Rock Dove *Columba livia*

Abundant resident.

Today, the feral pigeon is the most numerous bird in Kensington Gardens, but that has not always been so. A few censuses over the years have shown a rapid growth in numbers in the 1950s and 1960s, but an apparent levelling off since then:

1948	66
1966	620
1970	1,130
1995	969

The feral pigeons gather in flocks of a hundred or more, mainly by the park entrances where they wait for people to throw down sometimes quite large amounts of bread and grain for them. The practice of feeding the pigeons is officially discouraged. This is not easy to enforce since many people are quite protective towards them. Two ladies used regularly to catch injured birds, giving the pigeons a manicure after removing fishing line that had become entwined around their feet, while others talk to the birds, having pet names for some individuals.

Stock Dove *Columba oenas*

Uncommon breeding species.

It is probable that the Stock Dove is resident; it is certainly elusive even in summer. This most secretive of species has been breeding in Hyde Park and Kensington Gardens since at least the 1920s, but probably only as few as two or three pairs breed annually. In 1929 Macpherson noted that it was not infrequently seen and heard in Kensington Gardens, but he was not satisfied that it bred there then. A pair used to nest regularly some years before that in one of the elms near the bathing place in Hyde Park. At around 5 a.m. on 28th March 1951 as many as 25 were located in Kensington Gardens and seven pairs were found within a 200–300 yard radius of the Round Pond on 9th April 1954. There was a period in the 1960s with no firm evidence of breeding though birds were present. Two pairs were present during the 1994 Common Bird Census at potential nesting sites in Kensington Gardens. One was seen collecting small twigs in February 1995 at a new site near the Round Pond.

Winter records are rare; seven were seen in the December 1925 census and flying over in January 1926. During the 1970s there were reports of up to three birds in February in the years 1970, 1971 and 1975.

Woodpigeon *Columba palumbus*

Abundant resident.

The Woodpigeon was nesting in Kensington Gardens in 1878 and since 1884 had increased steadily. Macpherson thought there were four pairs in Kensington Gardens in 1886 (and perhaps as many more in the rest of Inner London). During the next half dozen years their increase was astounding and by 1891 it was quite common to see about 50 feeding together in Hyde Park. Their tameness increased as rapidly as their numbers. Macpherson reported counting over 100 roosting on a group of six trees by the Long Water on the evening of 21st February 1928 when many other trees in the neighbourhood were tenanted quite as thickly. The island in the Serpentine was another favourite roosting place. At the beginning of 1935 a vast number, estimated at several thousands, came into Kensington Gardens to roost. Most came from the north and west. Their numbers diminished rapidly towards the end of February. Several hundred which roosted on the island in the Serpentine arrived mostly from the east.

Numbers were very low in 1945, presumably because of extensive shooting during the Second World War following complaints by allotment holders. But today Woodpigeons are widespread throughout the parks and will come readily to the hand for food. Wintering numbers have certainly fallen since the 1925 and 1975 censuses in Kensington Gardens, when respective counts of 241 and 386 were made—a count on 6th February 1995 produced only 46.

Quite large movements of Woodpigeon take place over the capital in autumn and observers in Hyde Park and Kensington Gardens are well placed to watch them. In November 1969 counts of over 2,000 flying south-west were reported on three dates and that is probably typical of the direction and numbers to be seen in the early morning, but smaller movements in the opposite direction have sometimes been seen later in the day.

Collared Dove *Streptopelia decaocto*

Rare visitor.

Single pairs of this recent colonist have now bred in Inner London but the species has yet to colonise the parks and there are reports in only six years since an unconfirmed sighting on 3rd January 1964. The most recent report from the parks is of one on several dates in spring 1982.

Turtle Dove *Streptopelia turtur*

Uncommon passage migrant.

May is the most likely month for this bird to be seen, with records during the well-watched 1960s and 1970s in most years. In 1971 they were seen on 14 dates, including two sightings of up to five birds in September, and the following year on 11 dates in May and June. The most recent sighting was of one on 28th May 1987.

Cuckoo *Cuculus canorus*

Uncommon passage migrant.

Spring is the most obvious time for people to be aware of the Cuckoo and hardly a year went by without a report of at least one between the 1930s and the most recent report of one on 1st May 1986. As reports continue to come each year from Regent's Park, perhaps Cuckoos still come through Hyde Park and Kensington Gardens, but if so it is surprising that even the fewer observers of the parks during that period have missed hearing one for eight springs.

Barn Owl *Tyto alba*

Former breeder, not recorded this century.

Reduced to a single pair in 1897 and one or two in Kensington Gardens the following year. The closest and only report this century was of one at Hyde Park Corner flying into the Green Park on 11th May 1950.

Little Owl *Athene noctua*

Rare visitor with only five records this century.

One was seen perched in an elm in Kensington Gardens on 22nd April 1922, one was reported in Hyde Park in autumn 1925 and on 25th April 1936 Macpherson saw one in Kensington Palace Gardens. One was seen in Kensington Gardens on 10th October 1938 and twice subsequently that year. The most recent record is of one heard in Kensington Gardens on 21st July 1950.

Tawny Owl *Strix aluco*

Resident.

Prior to the tree losses from Dutch elm disease and the exceptional winds of 1987 and 1989 it was always possible to see roosting owls in the hollows of elm and horse-chestnut. But more recently, apart from a few daytime flights, they have become elusive. The last positive breeding record was in 1985. An estimate of four pairs in the two parks in the 1994 breeding season was based on birds heard calling during an authorised night visit in May. Nestboxes to attract this species were erected in 1994.

Short-eared Owl *Asio flammeus*

The only record is of one flying north-east over Kensington Gardens on 11th April 1974.

Nightjar *Caprimulgus europaeus*

Former rare passage visitor with only three records, none since 1922.

One was hawking over the bushes behind Apsley House, Hyde Park Corner, on 4th August 1908 and another was seen in the same neighbourhood on 19th July 1922. One was seen flying along the Serpentine near the bridge on 19th May 1921.

Swift *Apus apus*

Summer visitor and passage migrant.

A few are generally present feeding over the parks throughout the summer, but there are no known breeding sites nearby. Spring passage, involving up to 200 birds, was reported in 1981 and 1982.

Kingfisher *Alcedo atthis*

Rare visitor.

It is 1948 since two have been seen together and 1985 since even one was reported. Autumn used to be the most likely time to see one, with records in five of the years between 1977 and 1983.

Hoopoe *Upupa epops*

Rare visitor with only two records this century.

The observers who watched one flying round Kensington Gardens at lunchtime on 7th November 1967 regarded it as a sight they will always remember! In 1973 another bird was in Hyde Park, also on 7th November. In his *Birds of Great Britain* (1873), John Gould stated that one had been seen in Kensington Palace Gardens.

Wryneck *Jynx torquilla*

One was seen in the frames enclosure in Hyde Park on 9th April 1924. This is the only record this century; until 1850 it could be seen and heard in Kensington Gardens.

Green Woodpecker *Picus viridis*

Rare visitor.

Reported regularly in the 1940s and 1950s with the comment, in 1957, that the species must have suffered from the felling of diseased elm trees. That was a good prediction since the next record was not until 1969, when one was heard in April. One was heard on 8th August 1985 and one on 28th July 1991. In 1994 the frequency of records began to increase. There was one on 4th April and 14th July, during the Common Bird Census, and other reports during the latter part of the year and into early 1995. One observer thought that the bird had flown across from nearby Holland Park, where they are seen regularly.

Great Spotted Woodpecker *Dendrocopos major*

Uncommon visitor; has bred.

This species is most likely to be encountered in the parks in the autumn. Although is has been reported in the breeding season in many years since 1969, there has been no definite evidence of a nest since 1956. There were nine widely scattered records during the Common Bird Census in 1994, with a family party on 28th June. But whether they had nested in Kensington Gardens or elsewhere is uncertain. In 1995 there were sufficient records in the Buck Hill area and near the Hudson Memorial to indicate at least one territory.

Lesser Spotted Woodpecker *Dendrocopos minor*

Rare visitor.

Although birds were heard calling in Kensington Gardens before the Second World War and drumming was noted in May 1981 and 1985, the majority of the records received are of birds in autumn and winter. There were records of up to three birds in most years between 1973 and 1985.

Woodlark *Lullula arborea*

The only record is of one on the football pitches in Hyde Park on 15th January 1962.

Skylark *Alauda arvensis*

Passage migrant; only in hard weather are birds likely to be seen on the ground.

A favourite area used to be the Hyde Park football pitches and bowling green with their short grass. Up to 130 were counted in March 1970 and 76 for a few days in January 1971. More recently the only reports have referred to autumn passage of a few birds and to hard weather movements involving hundreds of birds, these movements also being widely observed throughout London.

Sand Martin *Riparia riparia*

Double passage migrant.

Almost certainly a regular passage migrant, but seldom staying for very long. Most records refer to up to 8 birds at a time, but in 1990 there were 10 on 3rd September and 50 on 4th September. Macpherson reported at least a hundred over the Serpentine and Long Water on 27th August 1896 and hundreds over the Long Water in late September and early October that year, with similar numbers over the Serpentine at the same time of year in 1899. Records seem evenly split between April and September, with isolated records in other months.

Swallow *Hirundo rustica*

Passage migrant.

More frequently reported in spring than autumn but numbers are small, usually only up to eight birds. An exception was in 1966 when numbers slowly built up to 50 between 13th and 22nd April, as the birds flew up and down the Serpentine in late snowstorms.

Autumn records are fewer but generally involve more birds. The years 1967, 1968 and 1976 had reports of more than 100 birds in the third week of September.

House Martin *Delichon urbica*

Summer resident and passage migrant.

Nesting began for the first time this century in 1978, when four nests were built under the eaves of a building by the Albert Gate entrance. This colony had grown to an estimated 36 nests by 1982, by which time they were also nesting on the French Embassy building on the east side of Albert Gate. Breeding success has fallen over much of the London Area and the size of the Hyde Park colony fell to eight nests by 1993. In 1994 scaffolding on one of the buildings confined nesting to the French

Embassy, where four pairs nested. The birds collect mud for their nests from puddles alongside the Serpentine and can be seen feeding over the water throughout the summer.

Spring passage is unspectacular but the return passage brings in more birds to join the local nesters feeding over the Serpentine. Over 200 were counted on 14th September 1976. More recently, 50 has been the largest number reported (in 1993).

Tree Pipit *Anthus trivialis*

Rare passage migrant.

In 1929 Macpherson reported that Tree Pipits were occasionally seen in spring in Kensington Gardens and Hyde Park; its visits were then rarer than they had been formerly. He saw one singing from the ground in Kensington Gardens on 10th May 1917. The species was stated to have been a regular spring visitor to Hyde Park between 1850 and 1860.

Between 1966 and 1981 there were records of at least one bird in all years except 1967 and 1973, but there has been none since 1981. There were spring records in 11 of the years (from mid April to early May) and autumn records in 10 of the years (between 6th August and 25th September), mostly of single birds.

Meadow Pipit *Anthus pratensis*

Passage migrant and occasional winter visitor.

A patient observer would probably see or hear one on passage on most mornings from mid March to mid April and from mid September to late October.[135] Numbers in spring have been up to eight birds, but in autumn from 'small numbers' up to a maximum of 200, on 5th October 1976, have been reported flying over. A few birds are occasionally put up from the banks of the Serpentine in winter. An exceptional record was of parties of 50 feeding on the newly-seeded underground car park in April 1964.

Rock Pipit *Anthus petrosus*

Rare visitor with only four records.

The first record for Kensington Gardens was of one feeding at the edge of the Round Pond, which was drained, on 5th January 1970. There was

---

135. W. G. Teagle saw Meadow Pipits *on the ground* round the Round Pond on just 4 of 140 visits between 8th August 1946 and 2nd January 1948; it was not until 1949 that London birdwatchers paid attention to the phenomenon of 'visible migration'.

another by the Round Pond on 4th October 1972 and singles in Hyde Park on 2nd January 1979 and 13th March 1980.

Yellow Wagtail *Motacilla flava*

Rare passage migrant.

In 1929 Macpherson reported that Yellow Wagtails were very rarely seen in spring in Kensington Gardens or Hyde Park, though they had been observed there in autumn more than once. One or two were seen on up to ten dates in all years between 1964 and 1983, with April, May, August and September being the favoured months. The recent decline in numbers nationally has been felt in the parks, with the most recent records being on 9th May 1986 and 28th April 1988.

Grey Wagtail *Motacilla cinerea*

Uncommon visitor; has bred nearby.

In 1929 Macpherson reported the Grey Wagtail as a fairly regular but not numerous visitor from August to February. Most recent records have been in the autumn, referring to one or two birds. Exceptional records are of seven on 4th October 1968 and four on 25th September 1980. The most recent record was in 1981, when one was seen on eight days between 26th August and 21st October. A favourite site used to be in the Italian Gardens when the fountain areas had been drained for cleaning.

In 1976 a pair seen collecting food near the Round Pond at the end of May nested just outside the south-west corner of Kensington Gardens. The following year a pair fed two fledged young by the Long Water on 1st June and had also probably nested just outside the park.

Pied Wagtail *Motacilla alba*

Resident; has bred.

One or two can usually be found feeding along the edge of the Serpentine, Long Water and Round Pond. In 1929 Macpherson reported that Pied Wagtails had bred on several occasions in Hyde Park and had done so at least once in Kensington Gardens. In the 1970s from one to three pairs appeared to have territories and bred inside or just outside the parks. Birds have been seen collecting food as recently as 1993, though during the Common Bird Census in 1994 there were only three records of single birds in April, June and July.

There are records of the Continental race *alba* ('White Wagtail'), all in spring, most recently on 25th April 1967.

Waxwing *Bombycilla garrulus*

The only record for the parks is of two adults and five immatures in Hyde Park on 30th January 1954.

Wren *Troglodytes troglodytes*

Common resident.

Usually found low down in most of the shrubberies, the Wren is often first located by its song and call. Breeding has taken place in every year since 1967. Before that there was some evidence of a reduced presence following severe weather and particularly after clearance work in the shrubberies; in 1952 it was stated to breed irregularly in Kensington Gardens and to be a non-breeding visitor to Hyde Park, while in the mid 1920s it was not a resident in either park, appearing in autumn. Common Bird Censuses were carried out between 1967 and 1980 and the average number of Wren territories held was 17. A similar census in 1994 produced 18 territories—9 in each park—pointing to a stable population.

Dunnock *Prunella modularis*

Common resident.

This retiring bird of the shrubberies is not well reported. The Common Bird Census results for 1967–1980 averaged 26 territories but a repeat census in 1994 found only 9, 4 in Hyde Park and 5 in Kensington Gardens.

Robin *Erithacus rubecula*

Common resident.

Now to be found in most shrubberies. The 1967–1980 Common Bird Censuses averaged 22 territories and a repeat census in 1994 totalled 18 territories, 9 in each park.

Nightingale *Luscinia megarhynchos*

Rare passage visitor with only two records this century.

Macpherson watched one in full song in Kensington Gardens early in the morning on 25th April 1936. The only other record is of one there on 1st May 1967. Nightingales were not infrequently heard in Hyde Park in the early part of the last century.

Black Redstart *Phoenicurus ochruros*

Rare visitor to the parks with only two records, both in 1969, of a female on 8th April and a male on 1st May.

Redstart *Phoenicurus phoenicurus*

Rare passage migrant; has bred.

In 1929 Macpherson reported that Redstarts had been regular in Kensington Gardens on spring migration but they had become much scarcer in the last 15 years. In 1891 he had described them as a regular summer migrant, passing through the Gardens every spring. At least one was seen each year during the 1970s, when the parks were well watched daily. During that time April, May and September produced the most sightings. There were four records of single birds in 1968 and 1971. The most recent record was of one on 2nd May 1981.

Two pairs are reported to have bred in Kensington Gardens in 1876.

Whinchat *Saxicola rubetra*

Rare passage migrant.

During the 1970s, when the parks were well watched, there were records in only five years. Sightings were in April, May or August. The most recent records are of two on 17th August 1978 and three on 8th May 1979. These were the only years with more than a single bird for almost 30 years.

Stonechat *Saxicola torquata*

Rare visitor with only four records this century.

One was seen by the Round Pond on 28th February 1912 and E. M. Nicholson saw one in the Long Water Sanctuary on 17th and 25th October 1925. An immature was in the sheepfold, long since removed, in Kensington Gardens on 14th September 1949 and one was in Hyde Park on 24th February 1977.

Wheatear *Oenanthe oenanthe*

Passage migrant.

Macpherson described the Wheatear as a regular spring visitor to the parks, less regular in autumn. On the morning of 8th April 1902 he saw about 200 in Hyde Park; they all left during the day.

More recently April, May and August have been the months with most records. One is seen in most years and during the 1970s up to six were seen together during the spring passage. In 1964, when the underground car park area had been newly seeded, there were 15 birds present on seven days between 7th April and 6th May and 10 birds on six days between 10th August and 7th September. One was reported near the police station in Hyde Park on 4th April 1994.

Birds identified as Greenland Wheatears *O. o. leucorrhoa* have been reported from time to time.

Ring Ouzel *Turdus torquatus*

Rare visitor on spring passage with only two records.

One was in Kensington Gardens on 28th April 1922 and another in the Dutch Garden near Kensington Palace on 19th May 1951.

Blackbird *Turdus merula*

Abundant resident.

The Blackbird is so familiar in the parks that it is seldom reported and only full and regular censuses can give a meaningful estimate of population levels and trends. There have been several autumn/winter censuses in Kensington Gardens:

1925–26 (average of three counts)	18
1948–49 (average of three counts)	91
1966–67 (average of three counts)	172
1975 (2nd November)	191

Common Bird Censuses were undertaken in both Hyde Park and Kensington Gardens during the period 1967–1980; unfortunately there is no sub-total available showing each park separately. The number of territories rose from 86 in 1967 to 150 in 1980. The rising trend was not entirely steady: there were setbacks in 1973 and again in 1979. A repeat census in 1994 produced an estimate of 60 territories (37 in Kensington Gardens), which represents a significant reduction since 1980. One difficulty in attempting to census Blackbirds results from the quite small territories which they hold in the shrubberies and among the trees, with apparently communal neutral feeding areas in the grassland nearby.

Macpherson heard one singing strongly in Kensington Gardens on 6th January 1903; he commented that it generally seems to acquire its song earlier in London than in the surrounding country. In 1891 he noted that a very pretty pied variety was to be seen near the Round Pond. Birds with white tail feathers and others with white collars could be found in the Dutch Garden in Kensington Palace Gardens and the nursery area in Hyde Park during 1994.

Fieldfare *Turdus pilaris*

Passage migrant and occasional winter visitor.

Autumn passage is reported in most years, with flocks of from 20 to 50 birds, exceptionally over 100, flying mostly north-west. Dates vary but are usually between mid October and late November. For example, in 1977 passage was observed between 28th October and 30th November with a maximum of 200 flying west on 18th November.

Spring passage is less well reported but a more gradual return seems to take place up until late April.

Parties occasionally stop to feed, usually during spells of cold weather. The largest feeding flock recorded in recent years was 370 on 6th March 1970. The most recent record is of six birds in Kensington Gardens on 21st and 24th February 1985.

Song Thrush *Turdus philomelos*

Common resident.

Results of the autumn/winter censuses of Kensington Gardens suggest that the species has never been numerous at that time of year:

1925–26 (average of three counts)	7
1948–49 (average of three counts)	5
1966–67 (average of three counts)	20
1975 (2nd November)	7

The Common Bird Censuses for the period 1967–1980, which were for both Hyde Park and Kensington Gardens, averaged 36 territories, within a range of 20 to 64 territories. The 1994 Common Bird Census showed only 9 territories, 5 of which were in Kensington Gardens. This represents a significant reduction since 1980. There was a less marked reduction in Blackbird numbers. During the 1967–1980 Censuses the ratio of Blackbird to Song Thrush averaged 4:1, whereas in the 1994 Census it was almost 7:1.

A comparison between the autumn counts and the much higher breeding numbers suggests that Song Thrushes leave the parks after nesting. The park thrushes might be vulnerable to the same problems as the rural population, which has declined in both farmland and woodland between 1970 and 1989.

Redwing *Turdus iliacus*

Passage migrant and winter visitor.

Autumn passage reports are similar to Fieldfare, except that they start in early October with a few records into December. In 1977 passage was reported from 11th October to 14th December with a maximum of 300 flying west on 18th November.

There are few spring records, up to 20 birds on 16 dates between January and March 1979 being the most recent. Feeding flocks of Redwings are much more likely to be encountered in the parks than Fieldfares. Their numbers and length of stay are influenced by the weather. In 1970 there were 317 counted on 6th March, during a cold spell, and 10 were still present on 10th April. Up to 20 were reported in November and December 1993 and flocks of this size are more usual.

Macpherson heard one in Kensington Gardens sing for a short time on 28th March 1917.

Mistle Thrush *Turdus viscivorus*

Resident.

In 1891 Macpherson said that a few pairs were resident and bred in Kensington Gardens. Numbers increased in winter. Although Mistle Thrushes were reported to have disappeared from Kensington Gardens in 1898, by 1929 they were nesting fairly regularly. They have done so in the parks since then and numbers have remained fairly constant since the 1960s. In 1994 there were five territories, only one of which was in Kensington Gardens. Mistle Thrushes are most likely to be found, throughout the year, on Buck Hill in Kensington Gardens and the area above the underground car park in Hyde Park.

Grasshopper Warbler *Locustella naevia*

Only two records, both in April.

One was reported reeling in the rabbit enclosure by the Long Water on 24th April 1934. The only other record is of one singing in Hyde Park on 13th April 1980.

Sedge Warbler *Acrocephalus schoenobaenus*

Rare passage migrant.

One or two birds are reported in April or May in most years. On 1st and 2nd May 1979 two birds were singing in the Long Water Sanctuary. There was an unusual record of one on 24th August 1970. One was reported from Kensington Gardens on 11th September 1935.

Marsh Warbler *Acrocephalus palustris*

The only record is of one seen and heard by H. G. Alexander in Kensington Gardens on 5th June 1924.

Reed Warbler *Acrocephalus scirpaceus*

Rare passage migrant.

Before 1928 the Reed Warbler was observed most years in Kensington Gardens, usually on spring passage, but rarely thereafter. More recently, there have been fewer records than for Sedge Warbler, with a maximum of one bird at a time, usually in May.

In 1891 Macpherson reported that this species had been recorded as nesting in Kensington Gardens. One was singing in the grounds of Kensington Palace on 21st June 1946, one was near the Long Water from 10th May to 28th June 1972 and one was in song near Kensington Palace on 31st May 1978. There were late August records in 1976 and 1981.

Lesser Whitethroat *Sylvia curruca*

Uncommon passage migrant; has bred.

One to three recorded in most years in May, with a scattering of records in August and September.

Nested in the framing-ground enclosure in Hyde Park in 1921. One sang frequently near the Long Water between 9th May and 2nd July 1942. A bird held territory in 1970 and a pair reared two young in 1978 near the Long Water.

Whitethroat *Sylvia communis*

Uncommon passage migrant.

One or two birds recorded in most years, with dates varying between April and May and August and September. Recent records are of one on 2nd September and two on 9th September 1980, one on 12th August 1981, one on 4th May and 13th August 1983 and one on 22nd and 23rd April 1987.

There is no proof of nesting but birds were reported carrying nesting material in 1925 (at the Kensington Gardens Sanctuary) and 1951.

Garden Warbler *Sylvia borin*

Uncommon passage migrant.

April, May, August and September are the most likely months in which to find this species, but only one to three birds are reported at any time.

Recent records are as follows: one on five dates between 1st and 14th May 1981; one on five dates between 27th July and 15th August 1981; three on 30th April 1982; singles on 5th, 7th and 11th May and 10th September 1982; one on 26th April 1983; one on 23rd August 1987; one on 15th September 1989; and one on 20th and 27th April 1990.

In 1915 one was singing strongly in Kensington Gardens as late as 29th July.

Blackcap *Sylvia atricapilla*

Passage migrant and summer resident.

A pair bred in 1961 but the eggs were taken. There was no further confirmed nesting until 1967, when there were three territories and at least one pair with young. They have probably nested each year since then, with a maximum of eight territories in 1972 and 1980. There were two territories in the Common Bird Census in 1994, including a nest in a private garden in Hyde Park.

A few passage birds are also seen in most years.

Wood Warbler *Phylloscopus sibilatrix*

Rare passage migrant.

In 1929 Macpherson reported that the song of the Wood Warbler was now generally heard once or twice each April or May in Kensington Gardens or Hyde Park; formerly such an occurrence was very rare. Recent records show how scarce and unpredictable are sightings now of this delightful species, with singles on 10th and 28th July 1967, one on 18th April 1968, one on 14th April 1969, singles on 17th July and 14th, 18th and 19th August 1975, one on 28th April 1978, one on 8th and two on 11th May 1979 and one in song on 26th April 1983.

Chiffchaff *Phylloscopus collybita*

Passage migrant; has bred.

On 15th September 1970 a total of 20 was counted, but more usually 3 is the maximum, at almost any time between late March and mid May and from late August to mid October, with some birds staying for several days.

In 1929 Macpherson reported that the Chiffchaff had been heard singing throughout the summer in Hyde Park. A territory was held in 1972 but there was no proof of breeding until 1989, when there were two territories.

There is one wintering record of a 'Willow/Chiff', which must surely have been a Chiffchaff, seen and heard on the Serpentine island on 20th December 1966.

Willow Warbler *Phylloscopus trochilus*

Common passage migrant; has bred.

A pair bred in Hyde Park in 1921 and 1922 and in Kensington Gardens in 1923. After a gap of 46 years, Willow Warblers nested successfully in 1969 and again in 1972. Territories were held up to 1977, but there has been no further proof of nesting.

Numbers on passage are sometimes quite impressive in both spring and autumn. Daily maxima are usually in the tens and twenties, with over 30 on 11th August 1971 and 17th August 1982. Most birds are reported between mid April and mid May and from late July to mid September.

Goldcrest *Regulus regulus*

Uncommon visitor.

Although five or more Goldcrests have been reported in every month of the year, at no time has there been a period of continuous observation that might suggest it is a resident species. During the 1970s, when the parks were well watched on a daily basis, the picture was far from clear. In 1976 one or two were seen between February and April and from September to November. In 1979 there were only three records in January and April of a single bird. The largest number reported was eight on 16th October 1981. More recently a territory was held in 1989 and 1990 in an area of Hyde Park where there is a small group of conifers. The species was not recorded during the 1994 Common Bird Census but birds were in evidence in early 1995 until at least 1st April.

Firecrest *Regulus ignicapillus*

Rare visitor.

The first records were in 1968, one in Hyde Park near *Rima* on 24th April and one in Kensington Gardens near *Peter Pan* on 21st October. One was in a holly tree on the west bank of the Long Water on 31st March 1970 and in bramble in the same area on 16th April. There were two records the following year, on 29th March and 24th September, and one in hawthorn on the west bank of the Long Water on 11th May 1973.

Spotted Flycatcher *Muscicapa striata*

Summer resident and passage migrant.

Has probably bred continuously in the parks for over a hundred years. In 1897 it was the only summer visitor still breeding and it was recorded as nesting regularly in 1909 and in 1929, when Macpherson reported that one or two pairs bred regularly in Kensington Gardens and Hyde Park; he recorded at least four pairs, possibly six, in 1893. In 1947 six nests were under observation in Kensington Gardens, but not all were successful; the species probably also bred in Hyde Park. During the Common Bird Census years, 1967–1980, there were between 3 and 15 territories mapped. However, only one territory was mapped in 1993 and 1994. In 1994 the birds nested in a private garden in Hyde Park but the eggs were predated.

In some years observers have reported an autumn passage. In the 1950s up to 20 were reported in the sheepfold which was a great attraction for many species. In the 1970s maxima of seven were described as passage birds between early August and, in some years, the end of September.

Pied Flycatcher *Ficedula hypoleuca*

Passage migrant, last recorded in 1983.

First recorded on 29th April 1890, the Pied Flycatcher was less than annual in and before the 1950s but for a time they became frequent on autumn migration. For example, in 1975 there were 17 on 16th August and the next year there were 100 'bird days' between 16th August and 30th September with a maximum of 10 birds on 3rd and 5th September. In 1983 there were only two on 11th August and three on 2nd September.

Long-tailed Tit *Aegithalos caudatus*

Resident.

The Long-tailed Tit was a mainly autumn visitor to the parks until 1972, when two pairs bred. Apart from a few years following the 1978/79 winter, they have bred regularly in small numbers since 1972. The 1994 Common Bird Census showed only one pair holding territory in Hyde Park but, on 28th February 1995, the remains of a nest were found in a gorse bush in Kensington Gardens and at least two pairs of Long-tailed Tits were exploring an area on the east side of the Long Water on the same day.

The largest flock reported is of 15 on 27th June 1984 but there were about 12 on 2nd November 1938.

Marsh Tit *Parus palustris*

Rare visitor. Four records involving five birds this century.

A pair was seen in Kensington Gardens on 10th February 1936. One was seen in company with other tits in Kensington Gardens on 4th and 28th October 1949 and one was seen in Hyde Park on 23rd March 1950. The most recent record is of one on 13th and 26th October 1966.

Willow Tit *Parus montanus*

The only record is of one near the Dell on 7th January 1972.

Coal Tit *Parus ater*

Resident.

They first bred in 1947 in Kensington Gardens, following widespread movements induced by a harsh winter, but not again until 1966. Since then they have nested regularly with a maximum of 16 territories and a minimum of 7 territories during the years of the regular Common Bird Census (1967–1980). The 1994 Common Bird Census showed 5 territories, 4 of which were in Kensington Gardens.

Blue Tit *Parus caeruleus*

Common resident.

The breeding population varied from 42 to 73 pairs during the years 1967–1980. In 1994 the number of territories was mapped at 35, of which 16 were in Kensington Gardens.

The species is widespread throughout both parks and many will come to the hand for food. They have used nestboxes, but one regular nesting site in Hyde Park is in the top of the Victorian lamp posts.

Great Tit *Parus major*

Common resident.

Not as numerous as the Blue Tit, the population estimate between 1967 and 1980 varied from 15 to 35 pairs. In 1994 there were 17 territories, 10 of them in Kensington Gardens.

This is another species which will come to the hand for food. They too have used nestboxes and some nested regularly in the base of Victorian lamp posts in Hyde Park, but these holes have recently been blocked up, possibly to stop them being used as rubbish receptacles.

Nuthatch *Sitta europaea*

Uncommon resident.

A sporadic nesting species with a maximum of two pairs reported. The first known breeding this century in Kensington Gardens was in 1958 following the arrival of the species the previous autumn in company with a widespread irruption of tits, though birds were heard calling in the breeding season before then and Yarrell recorded them as breeding in the Gardens in 1843. Apart from 1961, they nested each year until 1964, but after 1965 were reported only outside the breeding season until they nested again in 1976. One or two pairs have nested in most years since 1982, with the exception of 1988 and 1989. In the 1994 Common Bird Census there was one pair near the Kensington Gardens bandstand and an apparent unmated male calling near Speke's Monument.

Treecreeper *Certhia familiaris*

Uncommon resident.

Formerly an occasional visitor, first proved to breed this century in Kensington Gardens in 1945 and again in 1952 and 1954, after which it again became an occasional visitor. Now again probably a regular nesting species with up to four pairs reported during Common Bird Census work. Preferring the wooded areas of the parks, this is an elusive species. The most likely areas to find Treecreepers are in the chestnut trees on Buck Hill and among the oaks around the Kensington Gardens bandstand.

Red-backed Shrike *Lanius collurio*

One was seen in Hyde Park on 29th August 1904. This is the only record.

Jay *Garrulus glandarius*

Resident.

The Jay is most likely to be found outside the nesting season, particularly in autumn when family groups are about and the birds are conspicuously searching for acorns. During the nesting season many birds are believed to leave the parks to nest in nearby gardens. It has always been difficult to obtain proof of nesting, even though up to six territories were mapped during the 1967–1980 Common Bird Censuses. One exception was in 1975, when two dead nestlings were found near the Ranger's Lodge in Hyde Park. In 1994 there were only two sightings during the Common Bird Census, but up to five were seen together in September.

The Jay has not always been a resident species. Until 1928 it was only an occasional visitor to Inner London, although it was seen in Holland Park from 1923 onwards and was first proved to breed there in 1932. By 1938 the first pair bred in Hyde Park and in 1942 the first pair was proved to have nested in Kensington Gardens.

Magpie *Pica pica*

Resident.

Although Yarrell counted 28 Magpies together in Kensington Gardens (before 1843) and they bred there, probably for the last time, in 1856, they disappeared as a regular feature of the Gardens and were, for example, not recorded by E. M. Nicholson in 1925–1926 or at all between 1938 and 1949. The first recent attempts at nesting were in 1969 (when a nest was built in the Dell but the birds were driven off by Carrion Crows) and 1970. The first successful nest was in 1971 when two young were hatched in the Long Water Sanctuary. This nest and one in Regent's Park in the same season were the first breeding records for Inner London this century. The number of breeding pairs had risen to seven by 1980 and a similar total was recorded by the 1994 Common Bird Census.

In 1994 one bird in Hyde Park came to the hand for food.

Jackdaw *Corvus monedula*

A former resident, now irregular on passage.

The Jackdaw was a breeding species in small numbers until 1969. There was, for many years, a colony in the south-west corner of Kensington Gardens. There were fewer than a dozen pairs in Hudson's day and by 1929 most of the old trees in which the birds nested had either been blown down by gales or lopped for the safety of the public and, though several birds still frequented the area, it was doubtful whether any Jackdaws had bred there during the two previous years. A young bird being fed in Kensington Gardens on 16th June 1936 was the first to be reported from this small colony for some years. The last nesting area was in the elm trees in the south-west corner of the Gardens.[136] In 1948, when four pairs bred, one pair nested near the Orangery, a quarter of a mile to the north-west. Some of the birds were regularly fed on the balcony of a house just outside the park. The last resident birds were reported on 18th December 1970.

The autumn censuses of 1925 showed up to 8 birds present but they were most numerous in the 1940s, when up to 24 were reported in November 1945; 6 pairs were nesting from 1941 to 1947.

Since 1970 only birds flying over have been recorded with a maximum report of 10. Two birds stopped briefly in a tree on 9th October 1979 before flying off westwards. One flying over on 13th May 1974 was mobbed by Carrion Crows. The most recent report was of one on 24th January 1987.

---

136. A photograph of the site of the Jackdaw colony by C. B. Ashby appeared as plate 2a in *The Birds of the London Area since 1900* (R. C. Homes *et al.* 1957).

Rook *Corvus frugilegus*

Irregular on passage; formerly bred.

There were roughly 100 nests in Kensington Gardens, between the Broad Walk and the Serpentine, in 1836, diminishing to about 30 when, in 1880, the rookery was destroyed by the felling of 700 of the tallest elms in London. Hudson reported that even the very men who did the work respectfully informed the park authorities that they considered it a great mistake. There was a fresh but short-lived colony of about 12 nests in Kensington Gardens in 1893 and three nests in Hyde Park in 1900. The last known case of nesting in Inner London—in the Temple—was in 1916 as the distance to suitable feeding grounds increased.

A perhaps misguided attempt to reintroduce Rooks to Kensington Gardens was made in 1972. Four aviary bred birds were released initially, followed by more in 1973 and 1975. They attempted to nest in 1974 and 1975 but Carrion Crows forced them to desert. They were not seen after April 1976.

There have not been any records since one flew north-east on 7th March 1978. Prior to that there were a few mainly autumn and winter records of one to four birds overflying the parks.

Carrion Crow *Corvus corone*

Abundant resident.

In 1929 Macpherson described Kensington Gardens as the favourite haunt of the Carrion Crow in Inner London. Groups of more than 50 individuals feeding in Kensington Gardens have been reported since 1967. The autumn censuses show an increase from 5 in 1925 to 52 in 1975, despite 45 being shot in the latter year. The most recent count, on 6th February 1995, was 81, following further control of numbers during 1993 and 1994.

The Common Bird Census of Hyde Park and Kensington Gardens during the period 1967–1980 showed that up to 16 pairs were nesting in the two parks.

Many of the crows join groups of feral pigeons being fed by the public and, particularly by the Long Water, will approach very closely for food. Crows have been seen to take ducklings, feral pigeons and Starlings and a pair nesting in Kensington Palace Gardens in 1994 was blamed for the destruction of five nests, four of Blackbirds and one of Song Thrush.

A 'Hooded' Crow *C. c. cornix* was reported flying over on 16th February 1977.

Raven *Corvus corax*

A pair of Ravens bred annually on one of the large elms in Hyde Park until about 1826. One—probably an escape from the Tower of London—was seen near Kensington Palace in October 1889 and one remained in Kensington Gardens for some weeks in March and April 1890 until it was captured.

Starling *Sturnus vulgaris*

Common resident and passage migrant.

From being one of the most conspicuous birds in the parks in 1925, the Starling is now much less common. Numbers recorded have, however, been somewhat volatile, as illustrated by the following census figures for Kensington Gardens:

November 1925	411
December 1948	22
November 1966	82
November 1975	148
February 1995	43

The Common Bird Census of the 1970s included nests of Starlings. In 1968 there were 85 nests in Hyde Park and Kensington Gardens but by 1980 only 35 were mapped. Their major nest sites are the older trees with suitable holes. The exceptionally strong winds of some recent years and Dutch elm disease have both taken their toll of potential Starling nest sites.

Autumn passage is visible from the parks and huge flocks can sometimes be seen. In 1970 some 5,000 were counted on 22nd October and 1,500 flew west on 9th November 1976.

The island in the Serpentine was originally one of the Starling's favourite roosting-sites. By 1929 it and the trees in the Dell were regarded merely as suitable places for a rest before proceeding to some spot further east.

Some birds will settle on the hand for food, particularly in places where gulls are fed. Others have become adept at catching food thrown to the ducks before it hits the ground.

House Sparrow *Passer domesticus*

Common resident.

This species has suffered a most dramatic decline in numbers. A feeling that 'there are not as many sparrows now as there used to be' is confirmed by the following selection of census figures for Kensington Gardens:

November 1925	2,603
December 1948	885
November 1966	642
November 1975	544
February 1995	46

The number of potential nest sites has certainly declined. Tree holes, also used extensively by other species, are far fewer following the loss of many hundreds of the older trees to disease and strong winds and many old buildings, like the shelter on Buck Hill which used to have some 50 House Sparrow nests under the eaves, have been renovated. Evelyn Brown (*in litt.* 1994) has a vivid memory, probably from the 1930s, of a nesting colony of this species on a hawthorn tree on the edge of the bridge between the Serpentine and the Dell; she remembered it because of the 'weaver bird' characteristics of the nests. Unfortunately, the House Sparrow has not been included in most of the Common Bird Censuses but, on the basis of the census during the 1967 breeding season, the number of pairs of House Sparrows in Kensington Gardens was estimated at 160.

The reasons speculated for the national decline in House Sparrows—changes in farming practice and increasing numbers of predators, including Sparrowhawks—do not seem to apply to the parks unless the population, considered by some as largely sedentary, does require an injection of birds from outside London. The decline in horse traffic after 1925 has also been suggested as a possible factor.

There are a few places in the parks where House Sparrows will settle on the hand for food. They are certainly in competition for food with the feral pigeon. Some quite pale birds can be found and others with white wings and tails, giving some identification problems on occasion.

Tree Sparrow *Passer montanus*

Rare autumn and winter visitor.

Howard Saunders saw one in Kensington Gardens in 1899. Otherwise, almost all records refer to the period 1954 to 1977, with most referring to birds flying over. Autumn passage was reported in September or October in all those years, with a maximum in October 1970 of up to 22 birds on four dates. Mostly 1 to 10 birds were reported, with three or four sightings in the autumn of each year. From 1968 to 1970 there was a scattering of winter records with five flying south on 8th January 1968 and singles in January, February or March. There were spring records in 1969 and 1970 involving single birds on 14th and 18th April 1969 and 14th April 1970.

Chaffinch *Fringilla coelebs*

Mainly spring and summer visitor and passage migrant.

In 1929 Macpherson noted that during the previous few summers there must have been four or five pairs in Kensington Gardens and about as many in Hyde Park. The Chaffinch is still a regular breeding species with between 6 and 20 territories mapped in the two parks during the 1967–1980 Common Bird Censuses. In 1994 seven territories were held, perhaps surprisingly none of them in the Flower Walk.

Autumn and winter counts from Kensington Gardens confirm that most birds leave the park for the winter:

November 1925	10
December 1948	8
November 1966	1
November 1975	6
February 1995	0

Large numbers have been seen on autumn passage. In 1966 passage was apparent between 5th and 28th October, reaching a maximum of 420 flying north-west on the 11th. In 1968 passage from 7th to 17th October reached a maximum of 1,000 on the 15th.

Brambling *Fringilla montifringilla*

Rare double passage migrant; has occurred in winter.

Records cover both spring and autumn migration, mostly of four birds or fewer. An exception was 1975 when overhead passage was observed on nine dates between 19th October and 21st November with a maximum of seven on 22nd October. Bramblings were found in flocks of Chaffinches flying over on 6th March 1974 and 3rd October 1978. The most recent records are of singles on 12th March, 12th October and 8th November 1979.

Greenfinch *Carduelis chloris*

Resident and partial migrant.

Breeds regularly, with between 16 and 30 pairs in Hyde Park and Kensington Gardens estimated during the Common Bird Censuses in 1967–1980 but only 8 pairs mapped in 1994, 4 in each park. A few pairs were proved to breed in the late 1940s but the Greenfinch was absent as a breeding species for many years in the 1950s, as it was when Hudson wrote in 1898.

They are most in evidence in the breeding season and the autumn and winter censuses of Kensington Gardens confirm that most birds have left the parks by then:

November 1925	2
December 1948	0
November 1966	1
November 1975	10
February 1995	1

It is, however, not always like that. In 1977 a roost in Kensington Palace Gardens totalled over 550 birds on 29th November and 380 were counted in January 1978. In 1960 spent hops used for manure attracted up to 75 birds to the leaf yard, behind *Peter Pan*, during the latter half of January and early February.

Goldfinch *Carduelis carduelis*

Partial migrant; has nested nearby.

Most records refer to spring and autumn passage, with observations in April, May, late September and October. There are usually between 1 and 4 birds, but on 19th October 1977 there were 45 flying south, in 1978 passage was recorded from 20th September to 27th November, with a maximum of 15 flying south-west on 12th October, and 8 birds were seen on 26th April 1979.

There have been a few isolated winter records, most recently of from one to nine birds in January 1986. Birds have also been seen in the breeding season. On 8th July 1971 a pair was feeding three juveniles, in July 1972 young were being fed near the Victoria Gate, on 23rd July 1973 three juveniles were being fed, on 6th August 1976 there was a family group of four birds and on 17th August 1979 adults were feeding two juveniles. In each case the observer concluded that the birds had probably nested just outside the parks.

Siskin *Carduelis spinus*

Uncommon autumn and winter visitor.

The first records were as recently as 1968, when up to four birds were seen flying over on 10th January and 19th September. There are a few suitable areas of birch and alder where these birds may be found, in Hyde Park just north of the Serpentine and by the Long Water in Kensington Gardens, but they are insufficient to sustain a lengthy stay. Eight birds is the largest number reported, from 17th to 20th January 1986. The year with most records was 1974, with up to five on five days until 16th February, one on 10th April, singles on 9th, 22nd and 27th October and 3rd November, followed by three on 14th November. More typically there are only one to three records in most years, involving up to six birds. The most recent record is of two in Hyde Park from 11th to 23rd January 1989.

Linnet *Carduelis cannabina*

Passage migrant and uncommon winter visitor; may have bred.

Visible spring and autumn passage accounts for the majority of records. Observations are well spread out from March to May and from late September to early December. Numbers involved are generally in the range one to seven birds but larger flocks have occurred in autumn. On 5th October 1976 passage peaked with 200 flying south. The following year 35 flew south on 11th October and 34 flew south on 19th October.

A few birds have been seen feeding during hard weather. For example, the football pitches in Hyde Park attracted up to 10 birds during the winters of 1958 and 1962.

A pair frequented the framing-ground in Hyde Park during the summers of 1916 and 1918. There is some evidence that they bred there on the latter occasion.

Twite *Carduelis flavirostris*

E. M. Nicholson saw and heard two which were flying over the Round Pond on 21st November 1925 (see page 85). This is the only record and was the first for the London Area this century.

Redpoll *Carduelis flammea*

Passage migrant.

There were records of up to 20 birds feeding in birch trees in October–November in 1968, 1975 and 1978.

Birds have been seen flying over in all months except August, with records in most years from 1966 to 1983—there were no records in 1971 and 1972. The most recent records were in 1986 with 1 to 4 on several dates between 18th January and 2nd May, but there were 48 on 15th February.

Common Crossbill *Loxia curvirostra*

Three records involving birds flying over.

A party flew over Kensington Gardens on 21st November 1909. More recently, three flew over Kensington Gardens on 7th July 1966 (during a widespread irruption) and four flew north-west over Hyde Park, calling, on 15th August 1972 (during another irruption).

Bullfinch *Pyrrhula pyrrhula*

Uncommon resident.

Formerly an occasional visitor to Kensington Gardens and Hyde Park, the first evidence of nesting was in 1969 when two young were seen in August. One or two pairs have held territories in most years since then and nest-building was witnessed in 1985 and 1990. They are occasionally seen in the Hudson and Long Water Sanctuaries.

Hawfinch *Coccothraustes coccothraustes*

Rare visitor.

One was seen in Kensington Gardens on 25th April 1890. E. M. Nicholson saw two and heard the characteristic flight call over Rotten Row in Hyde Park at about 8.15 p.m. on 17th June 1930. One was in Kensington Gardens on 9th May 1939. In 1947 there were reports from the sheepfold in Kensington Gardens of a pair from 23rd April to 2nd May, a female on 21st April and between 8th and 25th July, and a male on 27th July. In *The Hawfinch* (1957), Guy Mountfort wrote that he had seen 'single birds or family parties during the last five years' in several parts of London, including both Kensington Gardens and Hyde Park. The latest dated records are of one on 19th November 1969 and one over the Round Pond on 15th October 1971.

Snow Bunting *Plectrophenax nivalis*

The only record is of one at the Round Pond on 26th November 1935.

Yellowhammer *Emberiza citrinella*

Uncommon visitor.

The only early records are of one singing by the bridge over the Serpentine on 5th June 1895, a female there on 5th January 1898, one seen and heard singing in Hyde Park on 24th April 1903, another seen there on 28th March 1928 and one with a flock of House Sparrows in Kensington Gardens in early December 1947.

Most subsequent records refer to one, or rarely two, birds flying over but there were seven in Hyde Park on 26th February 1958. February, March, April and October are the months with most records, but there were seldom more than three sightings in a year. The most recent records are of two on 16th and 21st February 1979 with one on the 22nd.

Reed Bunting *Emberiza schoeniclus*

Uncommon visitor.

All records are outside the breeding season, with as few as one or two records a year between 1962 and 1981.

INDEX

A

Alexander, H. G., 88, 116, 184
Ashby, C. B. (Brad), 21, 147, 191
Auk, Little, 172
Avocet, 165

B

Baker, Helen, 154
Ball, P., 128
Barn Elms Reservoirs, 50, 54
Barrie, J. M., 39
Battersea Park, 67
Beebe, C. William, 42, 126
Blackback, Greater. *See* Gull, Great Black-backed
Blackback, Lesser. *See* Gull, Lesser Black-backed
Blackbird, 75, 102, 120, 123, 130, 135, 182
Blackcap, 186
Boase, Henry, 15
Borrer, William, xiii
Boyd, Doug A., 128, 149
Brambling, 79, 85, 120, 131, 195
Brown, Evelyn P., 194
Brown, R. H., 37
Bullfinch, 131, 197
Bunting, Cirl, 143
Bunting, Reed, 198
Bunting, Snow, 198
Burkill, H. J., 47
Burkitt, J. P., 31, 116
Buzzard, Common, 162

C

Campbell, Bruce, 18
Chaffinch, 75, 102, 120, 123, 130, 194
Chance, Edgar P., 116
Chiffchaff, 77, 90, 91, 186
Church, Richard, 12, 39
Clark, R., 79

Committee on Bird Sanctuaries in the Royal Parks, 106, 128, 135, 138, 149
Coot, 77, 100, 124, 130, 164
Cormorant, 151
Corncrake, 164
Cornish, C. J., 51
Crake, Corn. *See* Corncrake
Cramp, Stanley, 61, 99, 128, 133, 149
Crossbill, Common, 197
Crow, Carrion, 75, 105, 120, 123, 130, 192
Crow, Hooded, 192
Cuckoo, 78, 98, 174
Curlew, 166

D

Dabchick. *See* Grebe, Little
Dancy, Eric, 110
Diver, Red-throated, 149
Dixon, Charles, xiii
Dove, Collared, 173
Dove, Rock (pigeon or London pigeon), 70, 131, 139, 172
Dove, Stock, 79, 84, 120, 131, 172
Dove, Turtle, 78, 92, 174
Duck, Ferruginous, 159
Duck, Long-tailed, 160
Duck, Mandarin, 131, 155
Duck, Tufted, 15, 70, 76, 120, 124, 130, 135, 159
Duck, Wild. *See* Mallard
Dunlin, 78, 84, 165
Dunnock, 76, 102, 120, 123, 130, 180

E

Earp, M. J. (Mike), 14
Epstein, Sir Jacob, 110

F

Falcon, Peregrine. *See* Peregrine
Fieldfare, 77, 84, 182
Firecrest, 187

Index

Fitter, R. S. R. (Richard), 71
Flycatcher, Pied, 78, 88, 188
Flycatcher, Spotted, 76, 89, 187
Forel, Professor, 127
Frampton, Sir George, 39
Frohawk, F. W., 24

G

Gadwall, 6, 76, 120, 130, 156
Gannet, 25, 151
Garganey, 158
Gibbons, David Wingfield, 84
Gibbs, A. (Tony), 83
Gillham, Eric, 18, 35
Glegg, William E., xiii, 73, 143, 149, 161, 164
Godwit, Bar-tailed, 166
Goldcrest, 130, 187
Goldeneye, 160
Goldfinch, 196
Goosander, 161
Goose, Brent, 155
Goose, Canada, 130, 131, 154
Goose, Greylag, 154
Goose, White-fronted, 154
Gordon, Seton, 20
Gould, John, 175
Grebe, Black-necked, 151
Grebe, Great Crested, 78, 97, 130, 150
Grebe, Horned. *See* Grebe, Slavonian
Grebe, Little (dabchick), 130, 149
Grebe, Red-necked, 150
Grebe, Slavonian, 150
Green Park, The, 101
Greenfinch, 76, 101, 120, 130, 195
Greenshank, 167
Guillemot, 171
Gull, Black-headed, 22, 35, 47, 77, 120, 124, 130, 134, 168
Gull, Common, 34, 44, 77, 120, 124, 130, 131, 169
Gull, Glaucous, 170
Gull, Great Black-backed (greater blackback), 21, 170
Gull, Herring, 21, 77, 130, 169
Gull, Iceland, 170
Gull, Lesser Black-backed (lesser blackback), 20, 78, 98, 130, 169

Gull, Little, 168

H

Harding, J. Rudge, 24, 35, 64, 79, 81
Harrier, Hen, 162
Harting, James Edmund, xiii, 143
Hawes, C. H., 149
Hawfinch, 198
Hayman, R. W., 47, 98, 149
Heron, Grey, 70, 78, 98, 115, 130, 152
Hobby, 163
Homes, R. C. (Dick), 191
Hoopoe, 175
Howard, Eliot, 31, 116
Hudson, W. H., xiii, 105, 110, 140, 149, 191, 192, 195
Hyde Park. *Passim*

J

Jackdaw, 75, 104, 120, 130, 191
Jay, 130, 190

K

Kensington Gardens. *Passim*
Kestrel, 77, 100, 120, 123, 130, 162
Kingfisher, 79, 175
Kite, Red, 161
Kittiwake, 170

L

Lack, David, 31
Lack, Elizabeth, 18
Lack, Lambert, 31
Lang, W. D., 99
Lapwing, 79, 84, 165
Lark, Sky. *See* Skylark
Lark, Wood. *See* Woodlark
Lebret, T., 10
Ledlie, R. C. B., 145
Linnet, 78, 86, 197
Littleton Reservoir. *See* Queen Mary Reservoir
Long Water, The (Kensington Gardens—same body of water as the Serpentine). *Passim*

Index

Low, George Carmichael, 147

M

Macpherson, A. Holte, 24, 66, 73, 79, 86, 147, 149, 152, 156, 157, 158, 164, 165, 167, 168, 169, 170, 171, 172, 173, 174, 177, 178, 179, 180, 181, 182, 184, 185, 186, 187, 192, 195
Magpie, 130, 190
Mallard (wild duck), 1, 7, 70, 76, 120, 124, 130, 134, 156
Mandarin. *See* Duck, Mandarin
Marchant, John H., 133
Martin, House, 78, 93, 177
Martin, Sand, 78, 93, 177
Massingham, H. J., 35, 55
McEwen, E., 149
Meinertzhagen, Richard, 35
Merganser, Red-breasted, 161
Merlin, 163
Moffat, Charles B., 31
Montier, David J., 70
Moorhen, 76, 112, 120, 124, 130, 164
Mountfort, Guy, 198

N

Newman, F., 79
Nicholson, Basil D. (EMN's brother), xiv, 8, 23, 33, 37, 52, 54, 60, 66, 72, 94, 96, 117, 127, 128
Nightingale, 180
Nightjar, 175
Nuthatch, 130, 189

O

O'Neill, Sir Con, 128
Oldham, Charles, 145
Oostveen, M. S. van, 147
Osborne, K. C. (Ken), 152
Osprey, 162
Ouzel, Ring, 182
Owl, Barn, 174
Owl, Little, 78, 98, 174
Owl, Short-eared, 175
Owl, Tawny, 98, 130, 174

Oystercatcher, 164

P

Parker, Eric, 64
Parrinder, E. R. (John), 21
Parsons, C. H. F., 149
Partridge, Grey, 163
Partridge, Red-legged, 163
Pedler, E. G., 145
Pennant, Thomas, 161
Peregrine, 163
Petrel, Leach's, 151
Petrel, Storm, 151
Pheasant, 163
Pigeon. *See* Dove, Rock
Pigeon, London. *See* Dove, Rock
Pigeon, Wood. *See* Woodpigeon
Piggott, Sir Thomas Digby, xiii
Pintail, 157
Pipit, Meadow, 78, 80, 178
Pipit, Rock, 178
Pipit, Tree, 178
Plover, Golden, 165
Plover, Ringed, 165
Pochard, 14, 70, 77, 120, 124, 130, 158
Pochard, Red-crested, 158
Pollard, Major, 98

Q

Queen Mary Reservoir, Littleton, 141

R

Rail, Water, 164
Raven, 193
Razorbill, 171
Redpoll, 130, 197
Redshank, 167
Redstart, 181
Redstart, Black, 180
Redwing, 77, 83, 120, 123, 130, 131, 183
Reid, James B., 84
Rhodes, Cecil, 12
Roberts, F. Russell, 79
Robin, 75, 102, 120, 123, 130, 180
Rook, 73, 131, 192

Index

Round Pond, The (Kensington Gardens). *Passim*
Russell, Harold, xiii, 79, 84, 100, 146

S

Sage, Bryan L., 35
Sanderling, 165
Sanderson, Roy F., 128, 133, 147, 149
Sandpiper, Common, 78, 92, 167
Sandpiper, Green, 167
Saunders, Howard, 194
Scaup, 160
Scoter, Common, 160
Serpentine, The (Hyde Park—same body of water as the Long Water). *Passim*
Severn, A. R., 79
Shag, 152
Shelduck, 155
Shoveler, 158
Shrike, Red-backed, 143, 190
Simms, Eric, 51
Siskin, 196
Skua, Arctic, 168
Skylark, 78, 80, 120, 130, 177
Smew, 161
Smith, Stuart, 99
Snipe, Common, 166
Sparrow, House, 75, 103, 120, 123, 124, 130, 131, 193
Sparrow, Tree, 194
Sparrowhawk, 79, 86, 130, 162
Speke, John Hanning, 32
Starling, 61, 72, 75, 103, 120, 123, 130, 134, 193
Stint, Little, 165
Stonechat, 78, 93, 181
Strangeman, Peter J., 151
Summers-Smith, J. Denis, 133
Swallow, 78, 92, 177
Swan, Bewick's, 153
Swan, Mute, 56, 130, 131, 153
Swan, Whooper, 153
Swann, Harry Kirke, xiii
Swift, 77, 99, 175

T

Teagle, W. G. (Bunny), 6, 73, 99, 128, 133, 134, 147, 149, 178
Teal, 156
Tern, Arctic, 171
Tern, Black, 171
Tern, Common, 171
Tern, Little, 171
Tern, Sandwich, 170
Terres, John K., 116
Throstle. *See* Thrush, Song
Thrush, Mistle, 76, 101, 120, 123, 130, 184
Thrush, Song (throstle), 75, 102, 120, 123, 130, 183
Tit, Blue, 75, 103, 120, 123, 130, 189
Tit, Coal, 77, 93, 120, 130, 189
Tit, Great, 75, 103, 120, 123, 130, 189
Tit, Long-tailed, 130, 188
Tit, Marsh, 188
Tit, Willow, 189
Tomlins, A. D., 147
Treecreeper, 130, 190
Twite, 78, 85, 88, 197

W

Wagtail, Grey, 79, 86, 179
Wagtail, Pied, 76, 100, 120, 123, 130, 179
Wagtail, White, 179
Wagtail, Yellow, 78, 87, 179
Wallace, D. I. M. (Ian), 83
Walthamstow Reservoirs, 54
Walton Reservoirs, 57
Warbler, Garden, 78, 88, 185
Warbler, Grasshopper, 184
Warbler, Marsh, 79, 88, 184
Warbler, Reed, 184
Warbler, Sedge, 78, 87, 184
Warbler, Willow, 76, 90, 187
Warbler, Wood (wood-wren), 78, 88, 91, 186
Watson, J. B., 79, 98
Watts, George Frederic, 12
Waxwing, 180
Wheatear, 181
Whimbrel, 166
Whinchat, 78, 87, 181

Index

Whitethroat, 77, 87, 99, 185
Whitethroat, Lesser, 185
Wigeon, 77, 155
Willow-wren. *See* Warbler, Willow
Woodcock, 166
Woodlark, 176
Woodpecker, Barred. *See* Woodpecker, Lesser Spotted
Woodpecker, Great Spotted (pied woodpecker), 77, 94, 120, 130, 176
Woodpecker, Green, 176
Woodpecker, Lesser Spotted (barred woodpecker), 77, 97, 176
Woodpecker, Pied. *See* Woodpecker, Great Spotted
Woodpigeon, 76, 111, 120, 123, 130, 134, 173
Wood-wren. *See* Warbler, Wood
Wren, 77, 100, 120, 130, 180
Wryneck, 176

Y

Yarrell, William, 75, 189, 191
Yellowhammer, 198

The London Natural History Society traces its history back to 1858 and today it is one of the largest societies of its kind in the world. It publishes a journal, *The London Naturalist*, and the *London Bird Report* annually together with a bimonthly *Newsletter* and *Ornithological Bulletin*, all of which are sent free to members. This is the Society's sixth book, having previously produced *The Birds of the London Area since 1900* (1957, revised 1964), *Atlas of the Breeding Birds of the London Area* (1977), *Flora of the London Area* (1983), *Butterflies of the London Area* (1987) and, most recently, *Larger Moths of the London Area* (1993). A new breeding bird atlas is under active preparation. The Society offers a full programme of indoor meetings and 'field' meetings, both 'turn-up' trips in and around London and organised coach trips further afield. A large selection of books and journals on most aspects of natural history is accessible to members at the Society's Library at South Kensington.

Applications to join the Society should be addressed to the LNHS Membership Secretary, *c/o* The Natural History Museum, Cromwell Road, London SW7 5BD. Enquiries about the Society's other publications should be addressed to the LNHS Publications Sales Secretary, *c/o* the same address.